GUIDE TO
ENVIRONMENTAL ANALYTICAL METHODS
2nd Edition

Edited by

Robert E. Wagner and William Kotas
Northeast Analytical, Inc.

Gregory A. Yogis
Environmental Consultant

Genium Publishing Corporation
One Genium Plaza, Schenectady, NY 12304-4690 (518) 377-8854

Copyright © 1992 by Genium Publishing Corporation
All Rights Reserved
ISBN 0-931690-44-7
Printed in the United States
2nd printing January, 1993
2nd Edition May, 1994

ACKNOWLEDGEMENTS

The original concept of an environmental analytical methods comparison was developed at the EPA Risk Reduction Engineering Laboratory (RREL) in Cincinnati, Ohio. The project, under the direction of Guy F. Simes, resulted in a publication titled *Variability in Protocols* (VIP). The goal of this publication was to help lab personnel compare various environmental analytical methods. However, in order to understand and effectively use VIP, the complete method needed to be at the user's fingertips. It also became quickly apparent that this information would be useful to people other than lab personnel performing the tests.

With the encouragement of the RREL, the information from *Variability in Protocols* was expanded by a commercial analytical lab, Northeast Analytical, Inc., Schenectady, New York, to include: (1) information needed by field personnel, (2) contacts for obtaining additional information about environmental analytical testing, and (3) a lexicon of terms and abbreviations. Additional information, such as the method tables referred to in the comparison and the listing of analytes, was added so that the user did not have to constantly refer back to the original documentation.

We would like to acknowledge and thank T. Chris Hynes, Sumiko Kiyonaga and other Northeast Environmental Laboratory members for the substantial assistance they gave researching, entering, sorting and reviewing the multitude of tables that went into making up this book.

GENERAL INTRODUCTION

The enormous amount of information that professionals in the environmental field have to deal with has become staggering. Volumes of federal publications fill our bookcases and also fill our time when using them. Primary reference materials for chemical analysis and associated analytical methodology are necessary for us to perform our jobs.

Often I have been asked the following questions. What analytes are covered by what method? What analytical methods are available and how do they differ from each other? What are the expected detection limits? What analysis method do you use for (name of a chemical or compound)? What type of container should be used? How much of a sample is required for analysis? Another set of questions seems to arise once the analysis is completed and the data are being reviewed. The questions are almost endless. The questions deserve to be answered, but it takes a tremendous amount of time to thumb through the regulations (methods) to find the information.

To help answer these questions and effectively deal with the never-ending task of methods comparison, the *Guide to Environmental Analytical Methods* has been prepared. Originally conceived by Guy F. Simes at the EPA Risk Reduction Engineering Laboratory in Cincinnati, Ohio and published as a booklet titled *Variability in Protocols,* the material has been expanded by Northeast Analytical, Inc. to become a stand-alone reference book for both lab and field personnel. The *Guide to Environmental Analytical Methods* condenses the information used most often from major environmental documentation such as SW-846; EPA 200, 500, and 600 Series; Standard Methods; and the Contract Laboratory Program (CLP) into one reference. The tabular approach effectively summarizes the key information (differences and similarities) of the most frequently referenced analytical methods. Included are method detection limits, calibration techniques, quality control requirements, analytes covered, sample storage, sample preservation, instrumentation, field sampling requirements, and other key parameters. There is also a compound cross-reference master list that allows the user to search for information by compound name. This table lists the methods in which the compound can be found, supplying an easy route for method selection. This eliminates the unpleasant task of searching through numerous methods to find the target compound, allowing easy searches from method to compound or compound to method in a condensed tabular format. To keep the guide as informative as possible, quality assurance/quality control tables from the original methods have been included in the Appendix along with vital EPA regional information and a glossary of common terms and abbreviations.

The *Guide to Environmental Analytical Methods* can be a valuable reference aid to many in the environmental field. Managers will find personnel spending less time searching for information for testing and client needs. Lab personnel will have method references, quality assurance documentation, and chemical compound information all in one source. Policy makers, auditors, QC officers, and personnel preparing quality assurance plans will find this document an extremely important reference. Instructors conducting environmental seminars will find the guide an excellent teaching aid. Also, field sampling crews will be able to carry the guide with them to the site. The *Guide to Environmental Analytical Methods* will give concise environmental chemical testing information and direct the individual back to the primary reference when the need arises.

Continued

INTRODUCTION TO 2nd EDITION

The response from the scientific community to the first edition of the *Guide to Environmental Analytical Methods* was overwhelming. This second edition contains our critical review of all method comparison sections included in the first edition, making the information current and up-to-date with published methodology. We have expanded information in many of the original topics and added clarity to several sections, including, for example, the QC Check Standards/Samples category. Additionally, the contact directory has been restructured and updated with several new information listings that provide the reader assistance with method interpretation.

Several methods that were included in the first edition have been replaced by newer versions. Replacements include the following: EPA method 525 is replaced with EPA method 525.1, SW-846 method 8020A is replaced with SW-846 method 8020, SW-846 method 8080A is replaced with SW-846 method 8080. Also SW-846 method 9010 has been removed from the cyanide comparison and is covered by SW-846 method 9010A.

Four entirely new methods have been added to the organic method comparison section to provide expanded coverage. The new methods are SW-846 method 8260, a GC/MS method for volatile organics; SW-846 method 8021, a GC method for volatile aromatic organics; EPA method 524.2, a GC/MS method for volatile organic compounds; and EPA method 502.2, a GC method for volatile organic compounds.

Everyone involved with the production of the *Guide to Environmental Analytical Methods* extends a thank you to those who have found "*The Guide*" a useful addition to their reference library.

Robert E. Wagner
Lab Director
Northeast Analytical, Inc.
301 Nott Street
Schenectady, NY 12305

Table of Contents

Analytical Methods - Organic Constituents

The numerous analytical methods for organic constituent testing have been organized into six major categories:

- Herbicides (chlorinated) identified by gas chromatography
- Pesticides (organochlorine) and PCBs identified by gas chromatography
- Semivolatile organic compounds identified by gas chromatography/mass spectroscopy
- Volatile organic compounds identified by gas chromatography/mass spectroscopy
- Volatile organic compounds (aromatic) identified by gas chromatography
- Volatile organic compounds (halogenated) identified by gas chromatography

Within each category there is a section called *Method Comparison* that summarizes the key method requirements and components for each testing method. The methods are compared on a side-by-side basis, which allows the user to quickly compare key aspects of the test methods. Where appropriate the user is referred to a table in Appendix A that provides details about an aspect of the test method. The tables in Appendix A are taken directly from the regulations.

For determining if a specific analyte is analyzed by a test method, each category has an *Analyte Listing* of all compounds addressed by the test methods within the category. This is also presented on a side-by-side basis so that the user can quickly see which method(s) cover the specific compound in question.

If you want to find a test method for a specific analyte but don't know what category(ies) it belongs to, refer to Appendix B. Appendix B is an alphabetical listing of all organic analytes specified in the regulations/methods included in this book. Use this appendix to identify the category to turn to for additional information.

Determination of Chlorinated Herbicides by GC

Method ⇒ Parameter ⇓	SW-846 Method 8150A	EPA 500 Series Method 515.1	Standard Methods Method 6640
Applicability	Groundwater, soils, sediments and waste samples	Groundwater, surface, and drinking water	Surface waters, groundwater
Number of Analytes (1)	10 total	17 total	Not specified (includes Silvex; 2,4-D; 2,4,5-T and "similar" chlorinated phenoxy acid chemicals).
Method Validation (2)	Extract and analyze 4 replicates of QC check standard. Compare accuracy and precision results to Table 3 (see page A-104).	Extract and analyze a minimum of 4 replicates of QC check sample. Results must be within ±30% of values in Table 2 (see page A-26).	Determine MDL (6)
QC Check Standards/ Samples	If MS/MSD results fall outside the ranges designated in Table 3 (see page A-104), a QC check standard must be analyzed and fall within those ranges.	Analyze one QC check sample (Laboratory Fortified Blank) per 20 samples or one per sample set (all samples extracted within a 24 hr period), whichever is more frequent. Concentration should be 10 times the EDL or MCL whichever is less. Compare %R to laboratory established limits, if available, or Table 2 (see page A-26). %R = R ± 30 or R ± 3 S_R whichever is broader.	Analyze control samples with each batch. Compare %R and %RSD to laboratory established limits.
Method Detection Limit	MDLs listed in Table 1 (see page A-102). EQLs listed in Table 2 (see page A-103).	EDLs listed in Table 2 (see page A-26). Periodically determine laboratory MDLs.	Varies with detector sensitivity, sample size, extract concentration.
Standard Solution Expiration (3)	Stock standards: 1 year Store at 4 °C in Teflon sealed vials, protect from light. Calibration standards: 6 months	Stock standards: 2 months Store at room temperature in Teflon sealed amber vials, protect from light. Calibration standards: not specified	Stock standards: 1 year Calibration standards: not specified
Initial Calibration	Minimum of 5 levels, lowest near but above MDL. If % RSD < 20, linearity assumed and average RF used. Alternatively use a calibration curve.	Minimum of 3 levels (5 recommended), lowest near but above EDL. If % RSD ≤ 20, linearity assumed and average RF used. Alternatively use a calibration curve or a single point calibration. Single point calibration standard must produce response within 20% of unknowns.	Prepare standard concentrations according to detector sensitivity and linearity. Number of standards used for calibration not specified.
Continuing Calibration	Mid-level calibration standard run every 10 samples. If not within ±15% of predicted response, recalibrate. Standard RT must fall within daily RT window or system is out of control. Samples injected after criteria was exceeded must be reanalyzed.	A calibration check standard must be analyzed at the beginning and end of analysis day (each at a different level). If not within ±20% of predicted response, recalibrate and reanalyze samples.	Inject same volume of herbicide methyl esters frequently to insure optimum operating conditions. Frequently is not defined in the method.
Surrogate Standards	One/two surrogates added to each sample (avoid use of deuterated analogs). Results must fall within laboratory established limits.	2,4-Dichlorophenylacetic acid (DCAA) at 0.5 µg/ml in extract. %R = 70 - 130.	Not specified.
Internal Standards	Optional: no standards specified.	Optional. If used, 4,4'-Dibromooctafluorobiphenyl (DBOB) at 0.25 µg/mL in extract.	Not specified.

Determination of Chlorinated Herbicides by GC (Continued)

Method ⇒ Parameter ⇓	SW-846 Method 8150A	EPA 500 Series Method 515.1	Standard Methods Method 6640
Accuracy/ Precision	One MS/MSD per 20 samples or each batch of samples, whichever is more frequent. Compare results to Table 3 (see page A-104). (See also **QC Check Standards/Samples.**)	One MS (Laboratory Fortified Sample Matrix) per 10 samples or each sample set whichever is more frequent. Compare %R to laboratory established limits. Analyze a QC sample from an external source at least quarterly. (See also: **QC Check Standards/ Samples.**)	One MS (Matrix with known additions) per 10 samples or at least one per month. Compare %R to laboratory established limits. Analyze one set of duplicates with each batch.
Blanks	One method blank per extraction batch (up to 20 samples) or when there is a change in reagents, whichever is more frequent.	One method blank with each batch of samples extracted or when new reagents used.	One method blank with each batch of samples.
Preservation/ Storage Conditions	Sodium thiosulfate if residual chlorine present. (aqueous) Store at 4 °C (aqueous). Store at 4 °C (solid).	Mercuric chloride as bactericide. Sodium thiosulfate if residual chlorine present. Protect from light. Store at 4 °C.	Store at 4 °C.
Holding Time (4)	Extraction: 7 days (aqueous) 14 days (solid) Analysis: 40 days after extraction	Extraction: 14 days Analysis: 28 days after extraction	Extraction: 7 days Analysis: 40 days after extraction
Field Sample Amount (5)	1 liter (aqueous) 4 oz. (solid) glass container Teflon lined top	1 liter glass container Teflon lined top	1 liter glass container Teflon lined top
Amount for Extraction	1 liter (aqueous) 50 grams (solid)	1 liter	850 to 1000 mL
Other Criteria (Method Specific)	When doubt exists in compound identification, second column or GC/MS confirmation should be used. QC for extraction and cleanup according to specific methods used.	Laboratory Performance Check Sample analyzed daily to monitor instrument sensitivity, column performance, and chromatographic performance. When doubt exists in compound identification, second column confirmation or additional qualitative technique must be used.	Use at least two columns for identification and quantification.

Notes:
(I) Analyte lists may vary among methods; a smaller list in one method is not necessarily a subset of a larger list in another method.
(2) Initial, one-time, demonstration of ability to generate acceptable accuracy and precision. Procedure may need to be repeated if changes in instrumentation or methodology occur.
(3) Indicates maximum usage time. If comparisons to QC check standards indicate a problem, more frequent preparation may be necessary.
(4) Unless otherwise indicated, holding times are from the date of sample collection.
(5) Approximate volumes to be gathered for analysis. Additional volumes are required for the generation of QC data.
(6) MDL determination for Standard Methods 6000 Series: Analyze a minimum of seven check samples (concentration = 0.2 times MCL or 10 times estimated MDL.). Average percent recovery should be 80% to 120% of true value with %RSD ≤ 35%. Use results to determine MDLs. Broader acceptance ranges exist for some compounds with lower extraction efficiency and are indicated in the specific method.

Determination of Chlorinated Herbicides by GC

Chemical Name	CAS Number	SW-846 Method 8150A	EPA 500 Series Method 515.1	Std. Methods Method 6640
Acifluorfen	50594-66-6		•	
Bentazon	25057-89-0		•	
Chloramben	133-90-4		•	
2,4-D	94-75-7	•	•	•
Dalapon	75-99-0	•	•	
2,4-DB	94-82-6	•	•	
DCPA acid metabolites	N/A		•	
Dicamba	1918-00-9	•	•	
3,5-Dichlorobenzoic acid	51-36-5		•	
Dichloroprop	120-36-5	•	•	
Dinoseb	88-85-7	•	•	
5-Hydroxydicamba	7600-50-2		•	
MCPA	94-74-6	•		
MCPP	93-65-2	•		
4-Nitrophenol	100-02-7		•	
Pentachlorophenol	87-86-5		•	
Picloram	1918-02-1		•	
2,4,5-T	93-76-5	•	•	•
2,4,5-TP(Silvex)	93-72-1	•	•	•

Determination of Organochlorine Pesticides and PCBs by GC

Method ⇒ Parameter ⇓	SW-846 Method 8080	EPA 500 Series Method 508	EPA 600 Series Method 608	Std. Methods Method 6630 (7)	CLP PEST (Organic SOW)
Applicability	Groundwater, soils, sludges, and non-water miscible wastes	Groundwater and finished drinking water	Municipal and industrial discharges	Drinking water, surface and groundwater, municipal and industrial discharges	Water, soil and sediment
Number of Analytes (1)	26 total	38 total (2 qualitative only)	25 total	35 listed (others possible).	28 total
Method Validation (2)	Extract and analyze 4 replicates of QC check standard. Compare accuracy and precision results to Table 3 (see page A-101).	Extract and analyze a minimum of 4 replicates of QC check sample. Results must be within ±30% of values specified in Table 2 (see page A-24).	Extract and analyze 4 replicates of QC check standards. Compare accuracy and precision results to Table 3 (see page A-53).	Extract and analyze 4 replicates of QC check standards. Compare accuracy and precision results to Table 6630:V (see page A-86). Determine MDL (6)	Not specified.
QC Check Standards/ Samples	If MS/MSD results fall outside ranges designated in Table 3 (see page A-101), a QC check standard must be analyzed and fall within those ranges.	Analyze one QC check sample (Laboratory Fortified Blank) per 20 samples or one per sample set (all samples extracted within a 24-hour period) whichever is more frequent. Concentration should be 10 times EDL or MCL whichever is less. Compare %R to laboratory established limits if available, or to Table 2 (see page A-24). %R = R ± 30 or R ± 3S$_R$ whichever is broader.	Analyze one QC check standard with every 10 samples. (Frequency may be reduced if MS recoveries meet QC criteria.) Compare %R to Table 3 (see page A-53).	Spike a minimum of 10% of all samples. If MS results fall outside ranges designated in Table 6630:V (see page A-86), a QC check standard must be analyzed and fall within those ranges.	Not specified.
Method Detection Limit	MDLs listed in Table 1 (see page A-99). PQLs listed in Table 2 (see page A-100).	EDLs listed in Table 2 (see page A-24). Periodically determine laboratory MDLs.	MDLs listed in Table 1 (see page A-52).	Varies with detector sensitivity, extraction/ clean-up efficiency, concentrations. MDLs listed in Table 6630:III (see page A-85).	CRQLs for TCL listed in Exhibit C (see page A-1).
Standard Solution Expiration (3)	Stock standards: 1 year. Store at 4 °C in Teflon sealed bottles, protect from light. Calibration standards: 6 months	Stock standards: 2 months. Store at room temperature in Teflon sealed amber vials, protect from light. Calibration standards: not specified	Stock standards: 6 months. Store at 4 °C in Teflon sealed bottles, protect from light. Calibration standards: not specified.	Stock standards: 6 months. Calibration standards: not specified.	Stock and calibration standards [except Performance Evaluation Mixture (PEM)] : 6 months. Store all standard solutions at 4 °C in Teflon sealed amber bottles, protect from light. PEM: 1 week.

Determination of Organochlorine Pesticides and PCBs by GC (Continued)

Method ⟹ Parameter ⇓	SW-846 Method 8080	EPA 500 Series Method 508	EPA 600 Series Method 608	Std. Methods Method 6630(7)	CLP PEST (Organic SOW)
Initial Calibration	Minimum of 5 levels, lowest near but above MDL. If %RSD < 20, linearity assumed and average RF used.	Minimum of 3 levels (5 recommended), lowest near but above EDL. If %RSD ≤ 20%, linearity assumed and average RF used.	Minimum of 3 levels, lowest near but above MDL. If %RSD <10, linearity assumed and average RF used.	Minimum of 3 levels, lowest near but above MDL. If %RSD <10, linearity assumed and average RF used. Alternatively, use a calibration curve. Calibrate system "daily."	Three levels (except multi-components at one level) the lowest at the CRQL. The %RSD must be ≤ 20% for target compounds, ≤ 30% for surrogates. Maximum of 2 target compounds may exceed 20 %RSD per column, but must be ≤ 30 %RSD. Analyze Resolution Check and PEM standards before initial calibration.
Continuing Calibration	Mid-level calibration standard run every 10 samples. If not within ± 15% of predicted response, recalibrate. Standard RT must fall within daily RT window or system is out of control. Samples injected after criteria was exceeded must be reanalyzed.	A calibration check standard must be analyzed at the beginning and end of analysis day (8 hrs.; each at a different level). If not within ± 20% of predicted response, recalibrate.	One or more calibration standards analyzed daily. If not within ±15% of predicted response, recalibrate.	Inject standards frequently to insure optimum operating conditions. Verify calibration each working day. If not within ±15% of predicted response, recalibrate.	Instrument Blank, Performance Evaluation Mixture and midpoint calibration standard mixtures A and B are run once per 12 hour period. Standard must be within ± 25% of predicted response.
Surrogate Standards	Two surrogates, dibutylchlorendate and 2,4,5,6- Tetrachloro-m-xylene. Results for one surrogate must fall within laboratory established control limits.	4,4'-Dichlorobiphenyl (DCB) or others if acceptance criteria met. %R = 70-130.	Not specified.	Not specified.	Two surrogates, Decachlorobiphenyl and 2,4,5,6-Tetrachloro-m-xylene. Advisory limits = 60-150%.
Internal Standards	Optional; no standards specified	Optional. If used, pentachloronitrobenzene or others if acceptance criteria met. Sample IS response must be ± 30% of daily standard response.	Optional; no standards specified.	Optional; not specified.	Not required.

GENIUM PUBLISHING CORPORATION

Determination of Organochlorine Pesticides and PCBs by GC (Continued)

Method ⇒ Parameter ⇓	SW-846 Method 8080	EPA 500 Series Method 508	EPA 600 Series Method 608	Std. Methods Method 6630(7)	CLP PEST (Organic SOW)
Accuracy/ Precision	One MS/MSD per 20 samples or each batch of samples, which- ever is more frequent. Compare results to Table 3 (see page A-101). (See also: **QC Check Standards/ Samples**)	One MS (Laboratory Fortified Sample Matrix) per 10 sam- ples or each set of samples, whichever is more frequent. Compare % R to lab- oratory established limits. Analyze a QC sample from an external source at least quarterly. (See also: **QC Check Standards/ Samples.**)	One MS per 10 samples from each site or 1 MS per month, whichever is more frequent. Compare %R to Table 3 (see page A-53). If MS results fall outside acceptance criteria, a QC check standard must be analyzed. (See also: **QC Check Standards/ Samples.**)	One MS per 10 samples from each site or 1 MS per month, whichever is more frequent. Compare %R to Table 6630:V (see page A-86). (See also: **QC Check Standards/ Samples.**)	One MS/MSD per sample delivery group or 1 per 20 samples, which- ever is greater. Six compounds are used for spiking; advisory limits in Exhibit D, Section III, Paragraph 16.4 (see page A-2).
Blanks	One method blank per extraction batch (up to 20 samples) or when there is a change in reagents, which- ever is more frequent.	One method blank with each batch of samples extracted or when new reagents used.	One method blank with each batch of samples extracted or when new reagents used.	One method blank with each batch of samples extracted.	One method blank per each case, or 20 samples (including MS and reanalyses) or each extraction batch or 14 day period samples received from case, which- ever is more fre- quent. Concentra- tion <CRQL for all target compounds. Instrument Blanks at start of 12 hr sequence, then every 12 hours. All target compounds <0.5 times CRQL.
Preservation/ Storage Conditions	If residual chlorine present, add sodium thiosulfate (aqueous) Store at 4 °C.	Mercuric chloride as bactericide. Sodium thiosulfate if residual chlorine present. Store at 4 °C.	Adjust pH to 5-9 if extraction not to be done within 72 hr of sampling. Add sodium thiosulfate if residual chlorine present and Aldrin is being determined. Store at 4 °C.	Adjust pH to 5-9 if extraction not to be done within 72 hr of sampling. Add sodium thiosulfate if residual chlorine present and Aldrin is being determin- ed. Store at 4 °C	Protect from light. Store at 4 °C
Holding Time (4)	Extraction: 7 days (aqueous) 14 days (solids) Analysis: 40 days after extraction	Extraction: 7 days Analysis: 14 days after extraction	Extraction: 7 days Analysis: 40 days after extraction	Extraction: 7 days Analysis: 40 days after extraction	Extraction: (separatory funnel) 5 days of sample receipt (aqueous) (Continuous liquid- liquid) started within 5 days of sample receipt (aqueous) 10 days of sample receipt (solid) Analysis: 40 days after extraction
Field Sample Amount Required (5)	1 liter (aqueous) 4 oz. (solid) glass container Teflon lined top	1 liter glass container Teflon lined top	1 liter glass container Teflon lined top	1 liter glass container Teflon lined top	1 liter (aqueous) 4 oz. (solid) glass container Teflon lined top

Determination of Organochlorine Pesticides and PCBs by GC (Continued)

Method ⇒ Parameter ⇓	SW-846 Method 8080	EPA 500 Series Method 508	EPA 600 Series Method 608	Std. Methods Method 6630(7)	CLP PEST (Organic SOW)
Amount for Extraction	1 liter (aqueous) 30 grams (low level solid) 2 grams (medium level solid)	1 liter	1 liter	1 liter	1 liter (aqueous) 30 grams (low level solid) 1 gram (medium level solid)
Other Criteria (Method Specific)	When doubt exists in compound identification, second column or GC/MS confirmation should be used. Check for dichloro-diphenyltrichloro-ethane (DDT) and endrin degradation. % breakdown must be ≤ 20%.	Laboratory Performance Check (LPC) sample analyzed daily to monitor instrument sensitivity, column performance, and chromatographic performance. Compare results to Table 3 (see page A-25) Check for dichloro-diphenyltrichloro-ethane (DDT) and endrin degradation. % breakdown must be ≤ 20%. At least quarterly analyze independent reference standard. Results must be within calibration check limits. When doubt exists in compound identification, second column confirmation should be used.	When doubt exists in compound identification, second column confirmation or GC/MS should be used.	When doubt exists in compound identification, second column confirmation or GC/MS should be used.	Second column confirmation is mandatory. GC/MS confirmation mandatory for positive samples of sufficient concentration. Check for dichloro-diphenyltrichloro-ethane (DDT) and endrin degradation every 12 hours. % breakdown must be < 20% for each compound and < 30% combined. Resolution criteria between adjacent peaks: mid-level individual standards ≥ 90%.; PEM = 100%.; Resolution Check Mixture ≥ 60%. A standard of an identified Aroclor must be run within 72 hrs of detection in a sample.

Notes:

(1) Analyte lists may vary among methods; a smaller list in one method is not necessarily a subset of a larger list in another method.

(2) Initial, one-time, demonstration of ability to generate acceptable accuracy and precision. Procedure may need to be repeated if changes in instrumentation or methodology occur.

(3) Indicates maximum usage time. If comparisons of QC check standards indicate a problem, more frequent preparation may be necessary.

(4) Unless otherwise indicated, holding times are from the date of sample collection.

(5) Approximate volumes to be gathered for analysis. Additional volumes are required for the generation of QC data.

(6) MDL determination for Standard Methods 6000 Series: Analyze a minimum of seven check samples (concentration = 0.2 times MCL or 10 times estimated MDL.). Average percent recovery should be 80% to 120% of true value with %RSD ≤ 35%. Use results to determine MDLs. Broader acceptance ranges exist for some compounds with lower extraction efficiency and are indicated in the specific method.

(7) Methods 6630B and 6630C are GC methods, Method 6630D is a GC/MS method cross referenced as Method 6410B.

Determination of Organochlorine Pesticides and PCBs by GC

Chemical Name	CAS Number	SW-846 Method 8080	EPA 500 Series Method 508	EPA 600 Series Method 608	Std. Methods Method 6630	CLP PEST (Organic SOW)
Aldrin	309-00-2	•	•	•	•	•
Aroclor-1016	12674-11-2	•	•	•	•	•
Aroclor-1221	11104-28-2	•	•	•	•	•
Aroclor-1232	11141-16-5	•	•	•	•	•
Aroclor-1242	53469-21-9	•	•	•	•	•
Aroclor-1248	12672-29-6	•	•	•	•	•
Aroclor-1254	11097-69-1	•	•	•	•	•
Aroclor-1260	11096-82-5	•	•	•	•	•
α-BHC	319-84-6	•	•	•	•	•
β-BHC	319-85-7	•	•	•	•	•
δ-BHC [1]	319-86-8	•	•	•	•	•
γ-BHC (Lindane)	58-89-9	•	•	•	•	•
Captan	133-06-2				•	
α-Chlordane	5103-71-9		•			•
Chlordane	57-74-9	•	•	•	•	
γ-Chlordane	5103-74-2		•			•
Chlorneb	2675-77-6		•			
Chlorobenzilate [1]	510-15-6		•			
Chlorothalonil	2921-88-2		•			
DCPA	1897-45-6		•			
4,4'-DDD	72-54-8	•	•	•	•	•
4,4'-DDE	72-55-9	•	•	•	•	•
4,4'-DDT	50-29-3	•	•	•	•	•
Dichloran	N/A				•	
Dieldrin	60-57-1	•	•	•	•	•
Endosulfan I	959-98-8	•	•	•	•	•
Endosulfan II	33213-65-9	•	•	•	•	•
Endosulfan sulfate	1031-07-8	•	•	•	•	•
Endrin	72-20-8	•	•	•	•	•
Endrin aldehyde	7421-93-4	•	•	•	•	•
Endrin ketone	53494-70-5					•
Etridiazole	2593-15-9		•			

Determination of Organochlorine Pesticides and PCBs by GC (Continued)

Chemical Name	CAS Number	SW-846 Method 8080	EPA 500 Series Method 508	EPA 600 Series Method 608	Std. Methods Method 6630	CLP PEST (Organic SOW)
Heptachlor	76-44-8	•	•	•	•	•
Heptachlor epoxide	1024-57-3	•	•	•	•	•
Hexachlorobenzene	118-74-1		•			
Malathion [2]	121-75-5				•	
Methoxychlor	72-43-5	•	•		•	•
Methyl parathion [2]	298-00-0				•	
Mirex	2385-85-5				•	
Parathion [2]	56-38-2				•	
Pentachloronitro-benzene (PCNB)	82-68-8				•	
cis-Permethrin	52645-53-1		•			
trans-Permethrin	52645-53-1		•			
Propachlor	1918-16-7		•			
Strobane	71-55-6				•	
Toxaphene	8001-35-2	•	•	•	•	•
Trifluralin [2]	1582-09-8		•		•	

Notes:

[1]   δ-BHC and Chlorobenzilate listed as tentatively measurable by Method 508.

[2]   Trifluralin and organophosphorus pesticides including Parathion, Methyl parathion and Malathion listed as tentatively measurable by Method 6630.

Determination of Semivolatile Organic Compounds by GC/MS

Method ⇒ Parameter ⇓	SW-846 Method 8270A	EPA 500 Series Method 525.1	EPA 600 Series Method 625	Std. Methods Method 6410	CLP SVOA (Organic SOW)
Applicability	Groundwaters, soils, sediments, sludges, and non-water miscible wastes	Drinking water and raw source water	Wastewater	Domestic, industrial wastewaters, natural and potable waters	Water, soil, and sediment
Number of Compounds (1)	233 total	43 total	81 total	81 total	64 total
Method Validation (2)	Extract and analyze 4 replicates of QC check standard. Compare accuracy and precision results to Table 6 (see page A-128).	Extract and analyze 4-7 replicates of QC check sample (Laboratory Fortified Blank). %R = 70-130 %RSD < 30 Use results to calculate MDLs. MDLs must be sufficient to detect analytes at regulatory levels.	Extract and analyze 4 replicates of QC check standard. Compare accuracy and precision results to Table 6 (see page A-61).	Extract and analyze 4 replicates of QC check standard. Compare accuracy and precision results to Table 6410:V (see page A-83). Determine MDL (6)	Not specified.
QC Check Standards/ Samples	If MS/MSD results fall outside ranges designated in Table 6 (see page A-128), a QC check standard must be analyzed and fall within those ranges.	Analyze one QC check sample (Laboratory Fortified Blank) per 20 samples or each batch of samples processed together within a work shift, whichever is more frequent. %R = 70-130 %RSD < 30 Use results to estimate MDLs. MDLs must be sufficient to detect analytes at regulatory levels Analyze replicate QC check samples and a QC sample from an external source at least quarterly.	If MS results fall outside ranges in Table 6 (see page A-61), a QC check standard must be analyzed and fall within those ranges.	If MS results fall outside ranges in Table 6410:V (see page A-83), a QC check standard must be analyzed and fall within those ranges.	Not specified.
Method Detection Limit	EQLs listed in Table 2 (see page A-121).	MDLs listed in Tables 2 (see page A-35), 3 (see page A-37), 4 (see page A-39), 5 (see page A-41), 6 (see page A-43), 7 (see page A-45), and 10 (see page A-46)	MDLs listed in Tables 4 (see page A-58) and 5 (see page A-60).	MDLs listed in Tables 6410:I (see page A-77) and 6410:II (see page A-80).	CRQLs for TCL listed in Exhibit C (see page A-3).

Determination of Semivolatile Organic Compounds by GC/MS (Continued)

Method ⇒ Parameter ⇓	SW-846 Method 8270A	EPA 500 Series Method 525.1	EPA 600 Series Method 625	Std. Methods Method 6410	CLP SVOA (Organic SOW)
Standard Solution Expiration (3)	Stock standards: 1 year. Store at 4 °C in Teflon sealed bottles, protect from light. Calibration standards: 1 year. Store between -10 °C to -20 °C. Daily continuing calibration standard: 1 week. Store at 4 °C	Stock standards: Expiration not specified. Store in dark, cool place in amber vials. Calibration standards: Expiration not specified. Store in dark, cool place in amber vials.	Stock standards: 6 months. Store at 4 °C in Teflon sealed bottles, protect from light. Calibration standards: not specified	Stock standards: 6 months. Store at 4 °C in TFE sealed bottles, protect from light. Calibration standards: not specified	Stock standards: 1 year. Store between -10 °C to -20 °C in Teflon sealed amber bottles. Calibration standards: 1 year. Store at -10 °C to -20 °C in Teflon sealed amber bottles. Daily continuing calibration standard: 1 week. Store at 4 °C (± 2 °C) in Teflon sealed amber bottles.
Initial Calibration	Minimum of 5 levels, lowest near but above MDL. %RSD for CCCs < 30. RF for SPCCs > 0.050.	Six levels. If %RSD < 30, average RF is used. Alternatively, use a calibration curve.	Minimum of 3 levels, lowest near but above MDL. If %RSD < 35, linearity assumed and average RF used. Alternatively, use a calibration curve.	Minimum of three levels, one near but above MDL. If % RSD < 35, linearity assumed and average RF used. Alternatively, use a calibration curve.	Five levels except for 8 compounds listed in **Other Criteria** which require 4 levels. RRF and %RSD criteria listed in Table 5 of Exhibit D/SV (see page A-6). Maximum of 4 compounds listed in Table 5 permitted to exceed criteria if RRF ≥ 0.010 and %RSD ≤ 40%. All others RRF ≥ 0.010.
Continuing Calibration	Mid-level calibration standard run every 12 hours. RF for SPCCs > 0.050. RF of CCCs must be < 30% difference from initial calibration.	Mid-level calibration standard analyzed at the beginning of each 8 hour shift. RF must be within ± 30% of initial calibration. Absolute areas of quant ions for IS and Surrogates must not decrease more than 30% from previous check standard, or by more than 50% from initial calibration.	One or more calibration standards analyzed each day. If not within ± 20% of predicted response, recalibrate.	One or more calibration standards analyzed each day. If not within ± 20% of predicted response, recalibrate.	A mid-level calibration standard is run every 12 hours. The % difference must be ≤ 25 from initial calibration. As with initial calibration up to 4 compounds in Table 5 of Exhibit D/SV (see page A-6) permitted to fail criteria if RRF > 0.010 and %D ≤ 40%.

Determination of Semivolatile Organic Compounds by GC/MS (Continued)

Method ⇒ Parameter ⇓	SW-846 Method 8270A	EPA 500 Series Method 525.1	EPA 600 Series Method 625	Standard Methods Method 6410	CLP SVOA (Organic SOW)
Surrogate Standards	Nitrobenzene-d$_5$, 2-Fluorobiphenyl, p-Terphenyl-d$_{14}$, Phenol-d$_6$, 2-Fluorophenol, and 2,4,6-Tribromophenol. Compare %R to laboratory established limits or to Table 8 (see page A-130) if laboratory limits unavailable. Laboratory limits must fall within limits listed in Table 8. (see page A-130)	Perylene-d$_{12}$. Optional: Caffeine-^{15}N$_2$ and Pyrene-d$_{10}$.	Minimum of three from Table 8 (see page A-63). % recovery limits not specified.	Minimum of 3 from Table 6410:IV (see page A-82). % recovery limits not specified.	Same as 8270A plus 1,2-Dichlorobenzene-d$_4$ and Chlorophenol-d$_4$. Recovery limits listed in Table 6 of Exhibit D/SV (see page A-8). If 2 or more acid surrogates or 2 or more BN surrogates fail criteria or %R < 10% for any surrogate, reanalyze/re-extract.
Internal Standards	1,4-Dichlorobenzene-d$_4$, Naphthalene-d$_8$, Acenaphthene-d$_{10}$, Phenanthrene-d$_{10}$, Chrysene-d$_{12}$, and Perylene-d$_{12}$. RT must be within ± 30 seconds from last calibration; area must be -50 to +100%.	Acenaphthene-d$_{10}$, Phenanthrene-d$_{10}$, and Chrysene-d$_{12}$. Sample IS response must be >70% of standard response.	Optional. If used, minimum of 3 from Table 8 (see page A-63).	Minimum of 3 [some recommended in Table 6410:IV (see page A-82)].	1,4-Dichlorobenzene-d$_4$, Naphthalene-d$_8$, Acenaphthene-d$_{10}$, Phenanthrene-d$_{10}$, Chrysene-d$_{12}$, and Perylene-d$_{12}$. IS RT must be within ± 30 seconds from last calibration; area must be -50 to +100%.
Accuracy/ Precision	One MS/MSD per 20 samples or each batch of samples, whichever is more frequent. Compare results to Table 6 (see page A-128). (See also: **QC Check Standards/ Samples.**)	Analyze MS replicates (Laboratory Fortified Sample Matrix) to determine effect of matrix. Analyze one MS per 20 samples if a variety of matrices are encountered. (See also: **QC Check Standards/ Samples.**)	One MS per 20 samples from each site or 1 MS per month, whichever is more frequent. Compare %R to Table 6 (see page A-61). (See also: **QC Check Standards/ Samples.**)	One MS per 20 samples from each site or 1 per month, whichever is more frequent. Compare results to Table 6410:V (see page A-83). (See also: **QC Check Standards/ Samples.**)	One MS/MSD per each case or 20 samples or each extraction batch or each 14 day period samples received from case, whichever is more frequent. 11 compounds used for spiking. Compare results to Table 7 of Exhibit D/SV (see page A-9).
Blanks	One method blank per extraction batch (up to 20 samples) or when there is a change in reagents, whichever is more frequent.	One method blank with each batch of samples extracted as a group within a work shift. Field Reagent Blank (FRB) is recommended with each sample set.	One method blank with each batch of samples extracted or when new reagents used.	One method blank with each batch of samples.	One method blank per each case, or 20 samples (including MS and reanalyses) or each extraction batch or grouped samples from case over 14 day period, whichever is more frequent. Concentration < CRQL for all compounds except phthalate esters < 5 times CRQL.

Determination of Semivolatile Organic Compounds by GC/MS (Continued)

Method ⇒ Parameter ⇓	SW-846 Method 8270A	EPA 500 Series Method 525.1	EPA 600 Series Method 625	Standard Methods Method 6410	CLP SVOA (Organic SOW)
Preservation/ Storage Conditions	Sodium thiosulfate if residual chlorine (aqueous). Store at 4 °C.	If residual chlorine, add sodium sulfite or sodium arsenite. Adjust pH ≤ 2 with HCl for unchlorinated water. Store at 4 °C.	Sodium thiosulfate if residual chlorine. Store at 4 °C.	Sodium thiosulfate if residual chlorine. Store at 4 °C.	Protect from light. Store at 4 °C.
Holding Time (4)	Extraction: 7 days (aqueous) 14 days (solids) Analysis: 40 days after extraction.	Extraction: 7 days Analysis: 30 days after extraction.	Extraction: 7 days Analysis: 40 days after extraction.	Extraction: 7 days Analysis: 40 days after extraction.	Extraction: (Continuous liquid-liquid) started within 5 days of sample receipt (aqueous);10 days of sample receipt (solid). Analysis: 40 days after extraction.
Field Sample Amount (5)	1 liter (aqueous) 4 oz. (solid) glass container Teflon lined top	1 liter glass container Teflon lined top	1 liter glass container Teflon lined top	1 liter glass container Teflon lined top	1 liter (aqueous) 4 oz. (solid) glass container Teflon lined top
Amount for Extraction	1 liter (aqueous) 30 grams (low level solid) 2 grams (medium level solid)	1 liter	1 liter	1 liter	1 liter (aqueous) 30 grams (low level solid) 1 gram (medium level solid)

Determination of Semivolatile Organic Compounds by GC/MS (Continued)

Method ⇒ Parameter ⇓	SW-846 Method 8270A	EPA 500 Series Method 525.1	EPA 600 Series Method 625	Standard Methods Method 6410	CLP SVOA (Organic SOW)
Other Criteria (Method Specific)	Tuning: 50 ng decafluorotriphenyl-phosphine (DFTPP) initially and every 12 hours; acceptance criteria in Table 3 (see page A-127). Qualitative ID: All ions > 10% intensity must be ± 20% of standard; ± 0.06 RRT units of standard RRT. Library searches may be made for the purpose of tentative identification.	Tuning: 5 ng deca-fluorotriphenylphos-phine (DFTPP) at the beginning of each 8 hour shift; acceptance criteria in Table 1 (see page A-34). Qualitative ID: All ions > 10% inten-sity must be ± 20% of standard ion; RT ± 10 seconds of standard RT. GC performance: Anthracene and phenanthrene should be separated by baseline. Benz[a]anthracene and crysene should be separated by a valley 25% of the average peak height. Greater than 99% of the compounds in the calibration standards must be recognized by peak identification software. The abundance of m/z 67 at the RT of endrin aldehyde must be <10% of the abundance of m/z 67 produced by endrin. LSE cartridges must be checked for contamination.	Tuning: 50 ng decafluorotri-phenylphosphine (DFTPP) at the beginning of each day; acceptance criteria in Table 9 (see page A-64). Qualitative ID: Characteristic ions for each analyte must maximize in the same scan or within 1 scan; rela-tive peak heights of 3 characteristic ions must be ± 20% of standard; RT (characteristic ions) within ± 30 seconds RT of analyte. Check column per-formance daily with benzidine and pentachlorophenol.	Tuning: 50 ng decafluorotriphenyl-phosphine (DFTPP) at the beginning of each day; accep-tance criteria in Table 6410:III (see page A-81). Qualitative ID: Characteristic masses [Table 6410:I (see page A-77) and Table 6410:II (see page A-80)] - each com-pound maximizes in same scan or within 1 scan; RT within ± 30 seconds of RT of standard; relative peak heights of 3 characteristic masses in EICPs within ± 20% of relative intensities of masses in refer-ence spectrum. Check column per-formance daily with benzidine and pentachlorophenol.	Tuning: 50 ng decafluorotriph-enylphosphine (DFTPP) initially and every 12 hours; acceptance criteria in Table 1 of Exhibit D (see page A-5). Up to 20 com-pounds with re-sponses >10% of nearest internal standard are ten-tatively identified via a library search. Qualitative ID: All ions > 10% inten-sity must be present and ± 20% of standard area; ± 0.06 RRT units of standard RRT. GPC cleanup of soils mandatory. Eight compounds, 2,4-Dinitrophenol, 2,4,5-Trichlorophenol, 2-Nitroaniline, 3-Nitroaniline, 4-Nitroaniline, 4-Nitrophenol, 4,6-Dinitro-2-methylphenol and Pentachlorophenol require only a 4 point calibration.

Notes:
(1) Analyte lists may vary among methods; a smaller list in one method is not necessarily a subset of a larger list in another method.
(2) Initial, one-time demonstration of ability to generate acceptable accuracy and precision. Procedure may need to be repeated if changes in instrumentation or methodology occur.
(3) Indicates maximum usage time. If comparisons to QC check standards indicate a problem, more frequent preparation may be necessary.
(4) Unless otherwise indicated, holding times are from the date of sample collection.
(5) Minimum volume for analysis. Additional volumes are required for the generation of QC data.
(6) MDL determination for Standard Methods 6000 Series: Analyze a minimum of seven check samples (concentration = 0.2 times MCL or 10 times estimated MDL.). Average percent recovery should be 80% to 120% of true value with %RSD ≤ 35%. Use results to determine MDLs. Broader acceptance ranges exist for some compounds with lower extraction efficiency and are indicated in the specific method.

Determination of Semivolatile Organic Compounds by GC/MS

Chemical Name	CAS Number	SW-846 Method 8270A	EPA 500 Series Method 525.1	EPA 600 Series Method 625	Std. Methods Method 6410	CLP SVOA (Organic SOW)
Acenaphthene	83-32-9	•		•	•	•
Acenaphthylene	208-96-8	•	•	•	•	•
Acetophenone	98-86-2	•				
1-Acetyl-2-thiourea	591-08-2	•				
2-Acetylaminofluorene	53-96-3	•				
Alachlor	15972-60-8		•			
Aldrin	309-00-2	•	•	•	•	
2-Aminoanthraquinone	117-79-3	•				
Aminoazobenzene	60-09-3	•				
4-Aminobiphenyl	92-67-1	•				
Anilazine	101-05-3	•				
Aniline	62-53-3	•				
o-Anisidine	90-04-0	•				
Anthracene	120-12-7	•	•	•	•	•
Aramite	140-57-8	•				
Aroclor-1016	12674-11-2	•		•	•	
Aroclor-1221	11104-28-2	•		•	•	
Aroclor-1232	11141-16-5	•		•	•	
Aroclor-1242	53469-21-9	•		•	•	
Aroclor-1248	12672-29-6	•		•	•	
Aroclor-1254	11097-69-1	•		•	•	
Aroclor-1260	11096-82-5	•		•	•	
Atrazine	1912-24-9		•			
Azinphos methyl	86-50-0	•				
Barban	101-27-9	•				
Benzidine	92-87-5	•		•	•	
Benzoic acid	65-85-0	•				
p-Benzoquinone	106-51-4	•				
Benzo(a)anthracene	56-55-3	•	•	•	•	•
Benzo(a)pyrene	50-32-8	•	•	•	•	•
Benzo(b)fluoranthene	205-99-2	•	•	•	•	•
Benzo(g,h,i)perylene	191-24-2	•	•	•	•	•
Benzo(k)fluoranthene	207-08-9	•	•	•	•	•
Benzyl alcohol	100-51-6	•				
α-BHC	319-84-6	•		•	•	
β-BHC	319-85-7	•		•	•	
δ-BHC	319-86-8	•		•	•	
γ-BHC (Lindane)	58-89-9	•	•	•	•	
4-Bromophenyl-phenylether	101-55-3	•		•	•	•
Bromoxynil	1689-84-5	•				
Butylbenzylphthalate[1]	85-68-7	•	•	•	•	•

[1]Some methods list as Benzyl butyl phthalate.

Determination of Semivolatile Organic Compounds by GC/MS (Continued)

Chemical Name	CAS Number	SW-846 Method 8270A	EPA 500 Series Method 525.1	EPA 600 Series Method 625	Std. Methods Method 6410	CLP SVOA (Organic SOW)
Captafol	2425-06-1	•				
Captan	133-06-2	•				
Carbaryl	63-25-2	•				
Carbazole	86-74-8					•
Carbofuran	1563-66-2	•				
Carbophenothion	786-19-6	•				
α-Chlordane	5103-71-9		•			
Chlordane	57-74-9	•		•	•	
γ-Chlordane	5103-74-2		•			
trans-Nonachlor-Chlordane	39765-80-5		•			
Chlorfenvinphos	470-90-6	•				
5-Chloro-2-methylaniline	95-79-4	•				
4-Chloro-3-methylphenol	59-50-7	•		•	•	•
4-Chloroaniline	106-47-8	•				•
Chlorobenzilate	510-15-6	•				
2-Chlorobiphenyl	2051-60-7		•			
bis(2-Chloroethoxy)methane	111-91-1	•		•	•	•
bis(2-Chloroethyl)ether	111-44-4	•		•	•	•
bis(2-Chloroisopropyl)ether	108-60-1	•		•	•	•
3-(Chloromethyl) pyridine hydrochloride	6959-48-4	•				
1-Chloronaphthalene	90-13-1	•				
2-Chloronaphthalene	91-58-7	•		•	•	•
2-Chlorophenol	95-57-8	•		•	•	•
4-Chlorophenyl-phenylether	7005-72-3	•		•	•	•
Chrysene	218-01-9	•	•	•	•	•
Coumaphos	56-72-4	•				
p-Cresidine	120-71-8	•				
Crotoxyphos	7700-17-6	•				
2-Cyclohexyl-4,6-dinitrophenol	131-89-5	•				
4,4'-DDD	72-54-8	•		•	•	
4,4'-DDE	72-55-9	•		•	•	
4,4'-DDT	50-29-3	•		•	•	
Demeton-o	298-03-3	•				
Demeton-s	126-75-0	•				
Diallate (*trans* or *cis*)	2303-16-4	•				
2,4-Diaminotoluene	95-80-7	•				
Dibenzofuran	132-64-9	•				•
Dibenzo(a,e)pyrene	192-65-4	•				
Dibenz(a,h)anthracene	53-70-3	•	•	•	•	•
Dibenz(a,j)acridine	224-42-0	•				

Determination of Semivolatile Organic Compounds by GC/MS (Continued)

Chemical Name	CAS Number	SW-846 Method 8270A	EPA 500 Series Method 525.1	EPA 600 Series Method 625	Std. Methods Method 6410	CLP SVOA (Organic SOW)
Dichlone	117-80-6	•				
1,2-Dichlorobenzene	95-50-1	•		•	•	•
1,3-Dichlorobenzene	541-73-1	•		•	•	•
1,4-Dichlorobenzene	106-46-7	•		•	•	•
3,3'-Dichlorobenzidine	91-94-1	•		•	•	•
2,3-Dichlorobiphenyl	16605-91-7		•			
2,4-Dichlorophenol	120-83-2	•		•	•	•
2,6-Dichlorophenol	87-65-0	•				
Dichlorovos	62-73-7	•				
Dicrotophos	141-66-2	•				
Dieldrin	60-57-1	•		•	•	
Diethyl sulfate	64-67-5	•				
Diethylphthalate	84-66-2	•	•	•	•	•
Diethylstilbesterol	56-53-1	•				
Dimethoate	60-51-5	•				
3,3'-Dimethoxybenzidine	119-90-4	•				
p-Dimethylaminoazobenzene	60-11-7	•				
3,3'-Dimethylbenzidine	119-93-7	•				
7,12-Dimethylbenz[A]anthracene	57-97-6	•				
2,4-Dimethylphenol	105-67-9	•		•	•	•
Dimethylphthalate	131-11-3	•	•	•	•	•
α,α-Dimethyphenethylamine	122-09-8	•				
Di-n-butylphthalate	84-74-2	•	•	•	•	•
4,6-Dinitro-2-methylphenol	121-14-2	•				•
1,3-Dinitrobenzene	99-65-0	•				
1,4-Dinitrobenzene	100-25-4	•				
1,2-Dinitrobenzene	528-29-0	•				
2,4-Dinitrophenol	51-28-5	•		•	•	•
2,4-Dinitrotoluene	121-14-2	•		•	•	•
2,6-Dinitrotoluene	606-20-2	•		•	•	•
Dinocap	6119-92-2	•				
Di-n-octylphthalate	117-84-0	•		•	•	•
Dinoseb	88-85-7	•				
Diphenylamine	122-39-4	•				
5,5-Diphenylhydantoin	57-41-0	•				
1,2 Diphenylhydrazine	122-66-7	•				
Disulfoton	298-04-4	•				
Di(2-ethylhexyl)adipate	103-23-1		•			
Endosulfan I	959-98-8	•		•	•	
Endosulfan II	33213-65-9	•		•	•	

Determination of Semivolatile Organic Compounds by GC/MS (Continued)

Chemical Name	CAS Number	SW-846 Method 8270A	EPA 500 Series Method 525.1	EPA 600 Series Method 625	Std. Methods Method 6410	CLP SVOA (Organic SOW)
Endosulfan sulfate	1031-07-8	•		•	•	
Endrin	72-20-8	•	•	•	•	
Endrin aldehyde	7421-93-4	•		•	•	
Endrin ketone	53494-70-5	•				
EPN	2104-64-5	•				
Ethion	563-12-2	•				
Ethyl carbamate	51-79-6	•				
Ethyl methanesulfonate	62-50-0	•				
bis(2-Ethylhexyl)phthalate	117-81-7	•	•	•	•	•
Famphur	52-85-7	•				
Fensulfothion	115-90-2	•				
Fenthion	55-38-9	•				
Fluchloralin	33245-39-5	•				
Fluoranthene	206-44-0	•		•	•	•
Fluorene	86-73-7	•	•	•	•	•
Heptachlor	76-44-8	•	•	•	•	
Heptachlor epoxide	1024-57-3	•	•	•	•	
2,2',3,3',4,4',6-Heptachlorobiphenyl	52663-71-5		•			
Hexachlorobenzene	118-74-1	•	•	•	•	•
2,2',4,4',5,6'-Hexachlorobiphenyl	60145-22-4		•			
Hexachlorobutadiene	87-68-3	•		•	•	•
Hexachlorocyclopentadiene	77-47-4	•	•	•	•	•
Hexachloroethane	67-72-1	•		•	•	•
Hexachlorophene	70-30-4	•				
Hexachloropropene	1888-71-7	•				
Hexamethyl phosphoramide	680-31-9	•				
Hydroquinone	123-31-9	•				
Indeno(1,2,3-cd)pyrene	193-39-5	•	•	•	•	•
Isodrin	465-73-6	•				
Isophorone	78-59-1	•		•	•	•
Isosafrole	120-58-1	•				
Kepone	143-50-0	•				
Leptophos	21609-90-5	•				
Maleic anhydride	108-31-6	•				
Malathion	121-75-5	•				
Mevinphos	7786-34-7	•				
Mestranol	72-33-3	•				
Methapyrilene	91-80-5	•				
Methoxychlor	72-43-5	•	•			

GENIUM PUBLISHING CORPORATION

Determination of Semivolatile Organic Compounds by GC/MS (Continued)

Chemical Name	CAS Number	SW-846 Method 8270A	EPA 500 Series Method 525.1	EPA 600 Series Method 625	Std. Methods Method 6410	CLP SVOA (Organic SOW)
Methyl parathion	298-00-0	•				
2-Methyl-4,6-dinitrophenol	534-52-1			•	•	
3-Methylcholanthrene	56-49-5	•				
4,4'-Methylenebis(2-chloroaniline)	101-14-4	•				
Methylmethanesulfonate	66-27-3	•				
2-Methylnaphthalene	91-57-6	•				•
2-Methylphenol	95-48-7	•				•
3-Methylphenol	108-39-4	•				
4-Methylphenol	106-44-5	•				•
Mexacarbate	315-18-4	•				
Mirex	2385-85-5	•				
Monocrotophos	6923-22-4	•				
Naled	300-76-5	•				
Naphthalene	91-20-3	•		•	•	•
1,4-Naphthoquinone	130-15-4	•				
2-Naphthylamine	91-59-8	•				
1-Naphtylamine	134-32-7	•				
Nicotine	54-11-5	•				
5-Nitroacenaphthene	602-87-9	•				
2-Nitroaniline	88-74-4	•				•
3-Nitroaniline	99-09-2	•				•
4-Nitroaniline	100-01-6	•				•
Nitrobenzene	98-95-3	•		•	•	•
4-Nitrobiphenyl	92-93-3	•				
Nitrofen	1836-75-5	•				
5-Nitro-o-anisidine	99-59-2	•				
5-Nitro-o-toluidine	99-55-8	•				
2-Nitrophenol	88-75-5	•		•	•	•
4-Nitrophenol	100-02-7	•		•	•	•
Nitroquinoline-1-oxide	56-57-5	•				
N-Nitrosodibutylamine	924-16-3	•				
N-Nitrosodiethylamine	55-18-5	•				
N-Nitroso-di-n-propylamine	621-64-7	•		•	•	•
N-Nitrosodiphenylamine	86-30-6	•		•	•	•
N-Nitrosodimethylamine	62-75-9	•		•	•	
N-Nitrosomethylethylamine	10595-95-6	•				
N-Nitrosomorpholine	59-89-2	•				
N-Nitrosopiperidine	100-75-4	•				
N-Nitrosopyrrolidine	930-55-2	•				
2,2',3,3',4,5',6,6'-Octachlorobiphenyl	40186-71-8		•			

Determination of Semivolatile Organic Compounds by GC/MS (Continued)

Chemical Name	CAS Number	SW-846 Method 8270A	EPA 500 Series Method 525.1	EPA 600 Series Method 625	Std. Methods Method 6410	CLP SVOA (Organic SOW)
Octamethyl pyrophosphoramide	152-16-9	•				
4,4'-Oxydianiline	101-80-4	•				
Parathion	56-38-2	•				
Pentachlorobenzene	608-93-5	•				
2,2',3',4,6-Pentachlorobiphenyl	60233-25-2		•			
Pentachloronitrobenzene (PCNB)	82-68-8	•				
Pentachlorophenol	87-86-5	•	•	•	•	•
Phenacetin	62-44-2	•				
Phenanthrene	85-01-8	•	•	•	•	•
Phenobarbital	50-06-6	•				
Phenol	108-95-2	•		•	•	•
1,4-Phenylenediamine	106-50-3	•				
Phorate	298-02-2	•				
Phosalone	2310-17-0	•				
Phosmet	732-11-6	•				
Phosphamidon	13171-21-6	•				
Phthalic anhydride	85-44-9	•				
2-Picoline	109-06-8	•				
Piperonyl sulfoxide	120-62-7	•				
Pronamide	23950-58-5	•				
Propylthiouracil	51-52-5	•				
Pyrene	129-00-0	•	•	•	•	•
Pyridine	110-86-1	•				
Resorcinol	108-46-3	•				
Safrole	94-59-7	•				
Simazine	122-34-9		•			
Strychnine	57-24-9					
Sulfallate	95-06-7	•				
Terbufos	13071-79-9	•				
1,2,4,5-Tetrachlorobenzene	95-94-3	•				
2,2',4,4'-Tetrachlorobiphenyl	2437-79-8		•			
2,3,4,6-Tetrachlorophenol	58-90-2	•				
Tetrachlorvinphos	961-11-5	•				
Tetraethyl pyrophosphate	107-49-3	•				
Thionazine	297-97-2	•				
Thiophenol (Benzenethiol)	108-98-5	•				
Toluene diisocyanate	584-84-9	•				
o-Toluidine	95-53-4	•				
Toxaphene	8001-35-2	•	•	•	•	

GENIUM PUBLISHING CORPORATION

Determination of Semivolatile Organic Compounds by GC/MS (Continued)

Chemical Name	CAS Number	SW-846 Method 8270A	EPA 500 Series Method 525.1	EPA 600 Series Method 625	Std. Methods Method 6410	CLP SVOA (Organic SOW)
1,2,4-Trichlorobenzene	120-82-1	•		•	•	•
2,4,5-Trichlorobiphenyl	15862-07-4		•			
2,4,5-Trichlorophenol	95-95-4	•				•
2,4,6-Trichlorophenol	88-06-2	•		•	•	•
0,0,0-Triethylphosphorothioate	126-68-1	•				
Trifluralin	1582-09-8	•				
Trimethyl phosphate	512-56-1	•				
2,4,5-Trimethylaniline	137-17-7	•				
1,3,5-Trinitrobenzene	99-35-4	•				
Tri-p-tolyl phosphate(h)	78-32-0	•				
Tris(2,3-dibromopropyl) phosphate	126-72-7	•				

Determination of Volatile Organic

Method ⇒ Parameter ⇓	SW-846 Method 8240A	SW-846 Method 8260	EPA 500 Series Method 524.1	EPA 500 Series Method 524.2
Applicability	Groundwaters. soils, sediments, sludges, non-water miscible wastes & others	Groundwaters. soils, sediments, sludges, non-water miscible wastes & others	Drinking water, raw source water	Drinking water, raw source water
Number of Analytes (I)	74 total	58 total	48 total	60 total
Method Validation (2)	Extract and analyze 4 replicates of QC check standard. Compare accuracy and precision results to Table 6 (see page A-108).	Extract and analyze 4 replicates of QC check standard. Compare to average recovery (R), standard deviation of recovery (S) and relative standard deviation (RSD) values given in Tables 7 (see page A-116) and 8 (see page A-118). Results for RSD must not exceed 2.6 times the single laboratory RSD or 20%, whichever is greater, and the mean recovery must lie within the interval R ± 3 times S or R ± 30%, whichever is greater.	Extract and analyze 4-7 replicates of QC check sample (Laboratory Fortified Blank). %R = 80-120 %RSD < 20. Use results to calculate MDLs. MDLs must be sufficient to detect analytes at regulatory levels.	Extract and analyze 4-7 replicates of QC check sample (Laboratory Fortified Blank). %R = 80-120 %RSD < 20. Use results to calculate MDLs. MDLs must be sufficient to detect analytes at regulatory levels.
QC Check Standards/ Samples	If MS/MSD results fall outside the ranges designated in Table 6 (see page A-108), a QC check standard must be analyzed and fall within those ranges.	The MS/MSD percent recovery (R_i) should lie within the QC acceptance criteria determined from the analysis of laboratory control samples during method validation. If the MS recovery is out of criteria, a QC check standard must be analyzed and fall within those ranges	Analyze one QC check sample (Laboratory Fortified Blank) per 20 samples or each batch of samples processed together within a work shift, whichever is more frequent. %R = 80-120 %RSD < 20 Use results to estimate MDLs. MDLs must be sufficient to detect analytes at regulatory levels.	Analyze one QC check sample (Laboratory Fortified Blank) per 20 samples or each batch of samples processed together within a work shift, whichever is more frequent. %R = 80-120 %RSD < 20 Use results to estimate MDLs. MDLs must be sufficient to detect analytes at regulatory levels.
Method Detection Limit	EQLs listed in Table 2 (see page A-105).	MDLs listed in Tables 1 (see page A-110) and 2 (see page A-112). EQLs listed in Table 3 (see page A-114)	MDLs listed in Table 3 (see page A-28).	MDLs listed in Tables 4 (see page A-30) and 5 (see page A-32).

Compounds by GC/MS

EPA 600 Series Method 624	Standard Methods Method 6210	CLP VOA (Organic SOW)	⇐ Method
			⇓ Parameter
Wastewater	Most environmental water samples	Water, soil, and sediment	**Applicability**
31 total	61 total	33 total	**Number of Analytes (I)**
Extract and analyze 4 replicates of QC check standard. Compare accuracy and precision results to Table 5 (see page A-57).	Analyze 4 replicates of QC check standard. Compare accuracy and precision results to Table 6210:IV (see page A-71). Determine MDL (6)	Not specified.	**Method Validation (2)**
Analyze one QC check standard per 10 samples. (Frequency may be reduced if MS recoveries meet QC criteria). Compare %R to Table 5 (See page A-57). If MS results fall outside acceptance criteria a QC check standard must be analyzed and fall within acceptance criteria.	Analyze a QC check standard with every 10 samples. (Frequency may be reduced if MS recoveries meet QC criteria). Compare %R to Table 6210:IV (see page A-71). If MS results fall outside acceptance criteria a QC check standard must be analyzed and fall within acceptance criteria.	Not specified.	**QC Check Standards**
MDLs listed in Table 1 (see page A-54).	MDLs listed in Table 6210:I (see page A-68).	CRQLs for TCL listed in Exhibit C (see page A-10).	**Method Detection Limit**

Determination of Volatile Organic

Method ⇒ Parameter ⇓	SW-846 Method 8240A	SW-846 Method 8260	EPA 500 Series Method 524.1	EPA 500 Series Method 524.2
Standard Solution Expiration (3)	Stock standards (except gases): 6 months. Store between -10 °C to -20 °C in Teflon sealed amber bottles with minimal headspace, protect from light. Stock gas standards: 2 months. Storage conditions same as stock standards. Calibration standards: Daily.	Stock standards (except gases): 6 months. Store between -10 °C to -20 °C in Teflon sealed amber bottles with minimal headspace, protect from light. Stock gas standards: 2 months. Storage conditions same as stock standards. Secondary dilution standards: 1 week. Storage conditions same as stock standards. Calibration standards: Daily	Stock and primary dilution standards (except gases): 4 weeks. Store at 4 °C in Teflon sealed bottles. Stock gas standards: 1 week. Store at < 0 °C or 1 day at room temperature. Aqueous calibration standards: 1 hour	Stock and primary dilution standards (except gases): 4 weeks. Store at 4 °C in Teflon sealed bottles. Stock gas standards: 1 week. Store at < 0 °C or 1 day at room temperature. Aqueous calibration standards: 1 hour
Initial Calibration	Minimum of 5 levels, lowest near but above MDL. %RSD for CCCs < 30. RF for SPCCs > 0.300 (> 0.250 for bromoform)	Minimum of 5 levels, lowest near but above MDL. %RSD for CCCs < 30. RF for SPCCs > 0.300 (> 0.250 for bromoform)	3, 4, or 5 levels (depending on calibration range), lowest 2-10 times MDL. If %RSD ≤ 20, linearity assumed and average RF used. Alternatively, use a calibration curve.	3, 4, or 5 levels (depending on calibration range), lowest 2-10 times MDL. If %RSD ≤ 20, linearity assumed and average RF used. Alternatively, use a calibration curve.
Continuing Calibration	Mid-level calibration standard run every 12 hours. RF for SPCCs same as for initial calibration. RF of CCCs must be < 25% difference from initial calibration.	Mid-level calibration standard run every 12 hours. RF for SPCCs same as for initial calibration. RF of CCCs must be < 25% difference from initial calibration.	Mid-level calibration standard analyzed at the beginning of each 8 hour shift. RF must be within ± 30% of initial calibration. Absolute areas of quant ions for IS and surrogates must not decrease more than 30% from previous check standard or by more than 50% from initial calibration.	Mid-level calibration standard analyzed at the beginning of each 8 hour shift. RF must be within ± 30% of initial calibration. Absolute areas of quant ions for IS and surrogates must not decrease more than 30% from previous check standard or by more than 50% from initial calibration.
Surrogate Standards	4-Bromofluorobenzene, 1,2-Dichloroethane-d_4, and Toluene-d_8. Compare %R to laboratory established limits or to Table 8 (see page A-109) if laboratory limits not available. Laboratory limits must fall within limits specified in Table 8.(see page A-109)	4-Bromofluorobenzene, dibromofluoromethane and Toluene-d_8. Compare %R to laboratory established limits or to Table 9 (see page A-120) if laboratory limits not available. Laboratory limits must fall within limits specified in Table 9. (see page A-120)	4-Bromofluorobenzene and 1,2-Dichlorobenzene -d_4. Additional surrogates optional. Recovery limits not specified. Absolute area of quant ions should not vary more than 50%.	4-Bromofluorobenzene and 1,2-Dichlorobenzene -d_4. Additional surrogates optional. Recovery limits not specified. Absolute area of quant ions should not vary more than 50%.
Internal Standards	Bromochloromethane, 1,4-difluorobenzene, and chlorobenzene-d_5. RT must be ± 30 seconds from last calibration; area must be 50 to 100%.	1,4-difluorobenzene, chlorobenzene-d_5, 1,4-dichlorobenzene-d_4 and pentafluorobenzene. RT must be ± 30 seconds from last calibration; area must be 50 to 100%.	Fluorobenzene. Additional internal standards optional.	Fluorobenzene. Additional internal standards optional.

Compounds by GC/MS   (Continued)

EPA 600 Series Method 624	Standard Methods Method 6210	CLP VOA (Organic SOW)	⇐ Method ⇓ Parameter
Stock standards (except gases): 1 month. Store between -10 °C and -20 °C in Teflon sealed bottles. Stock gas standards: 1 week. Stored between -10 °C and -20 °C in Teflon sealed bottles away from light. Aqueous calibration standards: 1 hour; 24 hours if stored at 4 °C in Teflon sealed bottles with zero headspace.	Stock standards (except gases and 2-chloroethyl-vinyl ether): 1 month. Store between -10 °C and -20 °C in Teflon sealed bottles away from light. Stock gas and 2-chloroethylvinyl ether standards: 1 week. Store between -10 °C and -20 °C. Aqueous calibration standards: 1 hour; 24 hours if stored in sealed vials with zero headspace.	Stock standards (except gases and reactive compounds): 6 months. Store between -10 °C and -20 °C in Teflon sealed vials with minimum headspace. Stock gas and reactive compounds standards: 2 months; store between -10 °C and -20 °C in Teflon sealed vials with minimum headspace. Aqueous calibration standards: 1 hour; 24 hours if stored at 4 °C in Teflon sealed vials with zero head space.	Standard Solution Expiration (3)
Minimum of 3 levels, lowest near but above MDL. If % RSD < 35, linearity assumed and average RF used. Alternatively, use a calibration curve.	Minimum of three levels, one near but above MDL. If %RSD < 35, linearity assumed and average RF used. Alternatively, use a calibration curve.	Five levels. RRF and RSD criteria listed in Table 2 of Exhibit D/VOA (see page A-12). Maximum of 2 compounds listed in Table 2 may exceed criteria if RRF ≥ 0.010 and %RSD ≤ 40%. All others target compounds RRF ≥ 0.010.	Initial Calibration
QC check standard analyzed each working day. Compare results to Table 5 (see page A-57).	QC check sample analyzed each working day and at a minimum frequency of 10% of all samples. Compare results to Table 6210:IV (see page A-71).	A mid-level calibration standard is analyzed every 12 hours. The % difference must be ≤ 25 from initial calibration. As with initial calibration, up to 2 compounds in Table 2 of Exhibit D/VOA (see page A-12) permitted to fail criteria if RRF > 0.010 and %D ≤ 40%.	Continuing Calibration
Minimum of 3 from Table 3 (see page A-56). % Recovery limits not specified.	Minimum of 3 from Table 6210:III (see page A-70). % Recovery limits not specified.	(System Monitoring Compounds) 4-Bromofluorobenzene, 1,2-Dichloroethane-d$_4$, and Toluene-d$_8$. Recovery limits in Table 6 of Exhibit D/VOA (see page A-13).	Surrogate Standards
Optional. If used, minimum of 3 from Table 3 (see page A-56).	Minimum of 3 from Table 6210:III (see page A-70).	Bromochloromethane, 1,4-difluorobenzene, and chlorobenzene-d$_5$. RT must be ± 30 seconds from last calibration; area must be 50 to 100%.	Internal Standards

Determination of Volatile Organic

Method ⇒ Parameter ⇓	SW-846 Method 8240A	SW-846 Method 8260	EPA 500 Series Method 524.1	EPA 500 Series Method 524.2
Accuracy/ Precision	One MS/MSD per 20 samples or each batch of samples, whichever is more frequent. Compare results to Table 6 (see page A-108). (See also: **QC Check Standards/Samples.**)	One MS/MSD per 20 samples or each batch of samples, whichever is more frequent. The MS/MSD percent recovery (R_i) should lie within the QC acceptance criteria determined from the analysis of laboratory control samples during method validation. (See also: **QC Check Standards/Samples.**)	No MS required. Replicate QC check standards and independent quality control sample (QCS) analyzed at least quarterly. (See also: **QC Check Standards/Samples.**)	No MS required. Replicate QC check standards and independent quality control sample (QCS) analyzed at least quarterly. (See also: **QC Check Standards/Samples.**)
Method Blanks	One method blank per extraction batch (up to 20 samples) or when there is a change in reagents, whichever is more frequent.	One method blank per extraction batch (up to 20 samples) or when there is a change in reagents, whichever is more frequent.	One method blank per 20 samples or batch of samples processed together, whichever is more frequent. One field reagent blank (FRB) per batch of samples.	One method blank per 20 samples or batch of samples processed together, whichever is more frequent. One field reagent blank (FRB) per batch of samples.
Preservation/ Storage Conditions	pH ≤ 2 with HCl or H_2SO_4 (aqueous). Sodium thiosulfate if residual chlorine (aqueous). Store at 4 °C.	pH ≤ 2 with HCl or H_2SO_4 (aqueous). Sodium thiosulfate if residual chlorine (aqueous). Store at 4 °C.	pH ≤ 2 with HCl. Ascorbic acid if residual chlorine. Collect all samples in duplicate. Store at 4 °C.	pH ≤ 2 with HCl. Ascorbic acid if residual chlorine. Collect all samples in duplicate. Store at 4 °C.
Holding Time (4)	14 days	14 days	14 days	14 days
Field Sample Amount Required (5)	40 mL VOA vial in duplicate without headspace (air bubbles) (aqueous) 4 oz. (solid) glass container Teflon lined septa	40 mL VOA vial in duplicate without headspace (air bubbles) (aqueous) 4 oz. (solid) glass container Teflon lined septa	40 mL VOA vial in duplicate without headspace (air bubbles) glass container Teflon lined septa	40 mL VOA vial in duplicate without headspace (air bubbles) glass container Teflon lined septa
Amount for Extraction	5 mL (aqueous) 5 grams (solid)	5 mL (aqueous) or 25 mL (aqueous) 5 grams (solid)	25 mL (5 mL may be used if MDLs can be achieved)	25 mL (5 mL may be used if MDLs can be achieved)

Compounds by GC/MS (Continued)

EPA 600 Series Method 624	Standard Methods Method 6210	CLP VOA (Organic SOW)	⇐ Method ⇓ Parameter
One MS per 20 samples from each site or 1 per month, whichever is more frequent. Compare %R to Table 5 (see page A-57). (See also: **QC Check Standards/Samples.**)	One MS per 10 samples analyzed and one MS per 20 samples from each site or one per month, whichever is more frequent. Compare results to Table 6210:IV (see page A-71). (See also: **QC Check Standards/Samples.**)	One MS/MSD per each case or 20 samples or each extraction batch or each 14 day period samples received from case, whichever is more frequent (only 5 compounds spiked). Compare results to Table 7 of Exhibit D/VOA (see page A-14).	**Accuracy/ Precision**
One method blank per day.	One method blank per day.	One method blank every 12 hours. Concentration ≤ CRQL for all but methylene chloride, acetone; 2-butanone ≤ 5 times CRQL.	**Method Blanks**
pH ≤ 2 with HCl. Sodium thiosulfate if residual chlorine. Store at 4 °C.	pH ≤ 2 with HCl. Sodium thiosulfate if residual chlorine. Store at 4 °C.	Protect from light. Store at 4 °C.	**Preservation/ Storage Conditions**
14 days	14 days	10 days from sample receipt	**Holding Time (4)**
40 mL VOA vial in duplicate without headspace (air bubbles) glass container Teflon lined septa	40 mL VOA vial in duplicate without headspace (air bubbles) glass container Teflon lined septa	40 mL VOA vial in duplicate without headspace (air bubbles) (aqueous) 4 oz. (solid) glass container Teflon lined septa	**Field Sample Amount Required (5)**
5 mL	5 mL	5 mL (aqueous) 5 grams (solid) 1 gram (medium-level solid)	**Amount for Extraction**

Determination of Volatile Organic

Method ⇒ Parameter ⇓	SW-846 Method 8240A	SW-846 Method 8260	EPA 500 Series Method 524.1	EPA 500 Series Method 524.2
Other Criteria (Method Specific)	Tuning: 50 ng bromo-fluorobenzene (BFB) initially and every 12 hours; acceptance criteria in Table 3 (see page A-107). Qualitative ID: All ions >10% intensity must be ± 20% of standard; ± 0.06 RRT units of standard RRT. Library searches may be made for the purpose of tentative identification.	Tuning: 50 ng bromofluorobenzene (BFB) initially and every 12 hours; acceptance criteria in Table 4 (see page A-115). Qualitative ID: All ions >10% intensity must be ± 20% of standard; ± 0.06 RRT units of standard RRT. Library searches may be made for the purpose of tentative identification. An analysis of a blank must follow a sample that has saturated ions from a compound.	Tuning: 50 ng bromofluorobenzene (BFB) at the beginning of each 8 hour shift; acceptance criteria in Table 2 (see page A-27). Qualitative ID: All ions >10% intensity must be ± 20% of standard ion; sample RT within 3 times SD of RT in calibration.	Tuning: 50 ng bromofluorobenzene (BFB) at the beginning of each 8 hour shift; acceptance criteria in Table 3 (see page A-29). Qualitative ID: All ions >10% intensity must be ± 20% of standard ion; sample RT within 3 times SD of RT in calibration.

Notes:

(l) Analyte list may vary among methods; a smaller list in one method is not necessarily a subset of a larger list in another method.

(2) Initial, one-time, demonstration of ability to generate acceptable accuracy and precision. Procedure may need to be repeated if changes in instrumentation or methodology occur.

(3) Indicates maximum usage time. If comparisons of QC check standards indicate a problem, more frequent preparation may be necessary.

(4) Unless otherwise indicated, holding times are from the date of sample collection.

(5) Approximate volumes to be gathered for analysis. Additional volumes are required for the generation of QC data.

(6) MDL determination for Standard Methods 6000 Series: Analyze a minimum of seven check samples (concentration = 0.2 times MCL or 10 times estimated MDL.). Average percent recovery should be 80% to 120% of true value with %RSD ≤ 35%. Use results to determine MDLs. Broader acceptance ranges exist for some compounds with lower extraction efficiency and are indicated in the specific method.

Compounds by GC/MS (Continued)

EPA 600 Series Method 624	Standard Methods Method 6210	CLP VOA (Organic SOW)	⇐ Method ⇓ Parameter
Tuning: 50 ng bromo-fluorobenzene (BFB) initially and every 12 hours; acceptance criteria in Table 2 (see page A-55). Qualitative ID: Three characteristic ions must be ± 20% of standard and must maximize within 1 scan.	Tuning: 50 ng bromo-fluorobenzene (BFB) at the beginning of each day; acceptance criteria in Table 6210:II (see page A-69). Qualitative ID: Characteristic masses of each compound must reach the maximization point in same scan or within 1 scan; RT within 30 seconds of RT of authentic compound; relative peak height of 3 characteristic masses in EICP within 20% of relative intensities of masses in reference spectrum.	Tuning: 50 ng bromo-fluorobenzene (BFB) initially and every 12 hours; acceptance criteria in Table 1 of Exhibit D/VOA (see page A-11). Qualitative ID: All ions >10% intensity must be present and ± 20% of standard; ± 0.06 RRT units of standard RRT. Up to 10 compounds with responses >10% of nearest IS are tentatively identified via a library search.	Other Criteria (Method Specific)

Determination of Volatile Organic Compounds by GC/MS

Chemical Name	CAS Number	SW-846 Method 8240A	SW-846 Method 8260	EPA 500 Series Method 524.1	EPA 500 Series Method 524.2	EPA 600 Series Method 624	Standard Methods Method 6210	CLP VOA (Organic SOW)
Acetone	67-64-1	•						•
Acetonitrile	75-05-8	•						
Acrolein	107-02-8	•						
Acrylonitrile	107-13-1	•						
Allyl alcohol	107-18-6	•						
Allyl chloride	107-05-1	•						
Benzene	71-43-2	•	•	•	•	•	•	•
Benzyl chloride (α-chlorotoluene)	100-44-7	•						
Bromoacetone	598-31-2	•						
Bromobenzene	108-86-1		•	•	•		•	
Bromochloromethane	74-97-5		•	•	•		•	
Bromodichloromethane	75-27-4	•	•	•	•	•	•	•
Bromoform	75-25-2	•	•	•	•	•	•	•
Bromomethane	74-83-9	•	•	•	•	•	•	•
2-Butanone	78-93-3	•						•
n-Butylbenzene	104-51-8		•	•			•	
sec-Butylbenzene	135-98-8		•	•			•	
tert-Butylbenzene	98-06-6		•	•			•	
Carbon disulfide	75-15-0	•						•
Carbon tetrachloride	56-23-5	•	•	•	•	•	•	•
Chlorobenzene	108-90-7	•	•	•	•	•	•	•
Chloroethane	75-00-3	•	•	•	•	•	•	•
2-Chloroethanol	107-07-3	•						
2-Chloroethylvinyl ether	110-75-8	•				•	•	
Chloroform	67-66-3	•	•	•	•	•	•	•
Chloromethane	74-87-3	•	•	•	•	•	•	•
Chloroprene	126-99-8	•						
3-Chloropropionitrile	542-76-7	•						
2-Chlorotoluene	95-49-8		•	•	•		•	
4-Chlorotoluene	106-43-4		•	•	•		•	
1,2-Dibromo-3-chloropropane	96-12-8	•	•	•	•		•	
Dibromochloromethane	124-48-1	•	•	•	•	•	•	•
1,2-Dibromoethane	106-93-4	•	•	•	•		•	
Dibromomethane	74-95-3	•	•	•	•		•	
1,4-Dichloro-2-butene	764-41-0	•						

Determination of Volatile Organic Compounds by GC/MS (Continued)

Chemical Name	CAS Number	SW-846 Method 8240A	SW-846 Method 8260	EPA 500 Series Method 524.1	EPA 500 Series Method 524.2	EPA 600 Series Method 624	Standard Methods Method 6210	CLP VOA (Organic SOW)
1,2-Dichlorobenzene	95-50-1		•	•	•	•	•	
1,3-Dichlorobenzene	541-73-1		•	•	•	•	•	
1,4-Dichlorobenzene	106-46-7		•	•	•	•	•	
Dichlorodifluoromethane	75-71-8	•	•	•	•		•	
1,1-Dichloroethane	75-34-3	•	•	•	•	•	•	•
1,2-Dichloroethane	107-06-2	•	•	•	•	•	•	•
1,1-Dichloroethene	75-35-4	•	•	•	•	•	•	•
cis-1,2-Dichloroethene	156-59-4		•	•	•		•	
trans-1,2-Dichloroethene	156-60-5	•	•	•	•	•	•	
1,2-Dichloroethene (total)	540-59-0							•
1,2-Dichloropropane	78-87-5	•	•	•	•	•	•	•
1,3-Dichloropropane	142-28-9		•	•	•		•	
2,2-Dichloropropane	590-20-7		•	•	•		•	
1,3 Dichloro-2-propanol	96-23-1	•						
1,1-Dichloropropene	563-58-6		•	•	•		•	
cis-1,3-Dichloropropene	10061-01-5	•		•	•	•	•	•
trans-1,3-Dichloropropene	10061-02-6	•		•	•	•	•	•
1,2,3,4 Diepoxybutane	298-18-0	•						
1,4 Dioxane	123-91-1	•						
Epichlorohydrin	106-89-8	•						
Ethanol	64-17-5	•						
Ethyl benzene	100-41-4	•	•	•	•	•	•	•
Ethyl methacrylate	97-63-2	•						
Ethylene oxide	75-21-8	•						
Hexachlorobutadiene	87-68-3		•		•		•	
2-Hexanone	591-78-6	•						•
2-Hydroxypropionitrile	78-97-7	•						
Iodomethane	74-88-4	•						
Isobutyl alcohol	78-83-1	•						
Isopropylbenzene	98-82-8		•		•		•	
4-Isopropyl toluene	99-87-6		•		•		•	
Malononitrile	109-77-3	•						
Methacrylonitrile	126-98-7	•						
Methyl iodide	74-88-4	•						
Methyl methacrylate	80-62-6	•						

Determination of Volatile Organic Compounds by GC/MS (Continued)

Chemical Name	CAS Number	SW-846 Method 8240	SW-846 Method 8260	EPA 500 Series Method 524.1	EPA 500 Series Method 524.2	EPA 600 Series Method 624	Standard Methods Method 6210	CLP VOA (Organic SOW)
4-Methyl-2-pentanone	108-10-1	•						•
Methylene chloride	75-09-2	•	•	•	•	•	•	•
Naphthalene	91-20-3		•		•		•	
Pentachloroethane	76-01-7	•						
2-Picoline	109-06-8	•						
Propargyl alcohol	107-19-7	•						
6-Propiolactone	57-57-8	•						
Propionitrile	107-12-0	•						
n-Propylamine	107-10-8	•						
n-Propylbenzene	103-65-1		•		•		•	
Pyridine	110-86-1	•						
Styrene	100-42-5	•	•	•	•		•	•
1,1,1,2-Tetrachloroethane	630-20-6	•	•	•	•		•	
1,1,2,2-Tetrachloroethane	79-34-5	•	•	•	•	•	•	
Tetrachloroethene	127-18-4	•	•	•	•	•	•	•
Toluene	108-88-3	•	•	•	•	•	•	•
1,2,3-Trichlorobenzene	87-61-6		•		•		•	
1,2,4-Trichlorobenzene	120-82-1		•		•		•	
1,1,1-Trichloroethane	71-55-6	•	•	•	•	•	•	•
1,1,2-Trichloroethane	79-00-5	•	•	•	•	•	•	•
Trichloroethene	79-01-6	•	•	•	•	•	•	•
Trichlorofluoromethane	75-69-4	•	•	•	•	•	•	
1,2,3-Trichloropropane	96-18-4	•	•	•	•		•	
1,2,4-Trimethylbenzene	95-63-6		•		•		•	
1,3,5-Trimethylbenzene	108-67-8		•		•		•	
Vinyl acetate	108-05-4	•						
Vinyl chloride	75-01-4	•	•	•	•	•	•	•
m-Xylene	108-38-3		•	•	•		•	
o-Xylene	95-47-6		•	•	•		•	
p-Xylene	106-42-3		•	•	•		•	
Xylene (total)	1330-20-7	•						•

Determination of Aromatic Volatile Organic Compounds by GC

Method ⇒ Parameter ⇓	SW-846 Method 8020	EPA 500 Series Method 503.1	EPA 600 Series Method 602	Standard Methods Method 6220
Applicability	Groundwater, soils, sludges, water-miscible and non-water miscible wastes	Drinking water, raw source water	Municipal and industrial discharge water	Most environmental water samples
Number of Analytes (1)	8 total	28 total	7 total (same as 6220)	7 total (same as 602)
Method Validation (2)	Analyze 4 replicates of QC check standard. Compare accuracy and precision results to Table 3 (see page A-93).	Analyze 4-7 replicates of QC check sample (Laboratory Fortified Blank). %R = 80-120. %RSD < 20. Calculate MDLs, results must be sufficient to detect analytes at required levels.	Analyze 4 replicates of QC check standard. Compare accuracy and precision results to Table 2 (see page A-51).	Analyze 4 replicates of QC check standard. Compare accuracy and precision results to Table 6220:II (see page A-73). Determine MDL(6)
QC Check Standards/ Samples	If MS/MSD results fall outside ranges designated in Table 3 (see page A-93), a QC check standard must be analyzed and fall within those ranges.	Analyze one QC check sample (Laboratory Fortified Blank) per 20 samples or each batch of samples processed together within a work shift, whichever is more frequent. %R = 80-120 %RSD < 20 Use results to estimate MDLs. MDLs must be sufficient to detect analytes at regulatory levels. Analyze replicate QC check samples and a QC sample from an external source at least quarterly.	If MS results fall outside ranges designated in Table 2 (see page A-51), a QC check standard must be analyzed and fall within those ranges.	If MS results fall outside ranges designated in Table 6220:II (see page A-73), a QC check standard must be analyzed and fall within those ranges.
Method Detection Limit	MDLs listed in Table 1 (see page A-92).	MDLs listed in Table 2 (see page A-23).	MDLs listed in Table 1 (see page A-50).	MDLs listed in Table 6220:I (see page A-72).
Standard Solution Expiration (3)	Stock standards: 6 months. Store at 4 °C in Teflon sealed bottles with minimum headspace, away from light. Aqueous calibration standards: 1 hour; 24 hours if stored with zero headspace.	Stock and primary dilution standards: 4 weeks. Store at 4 °C in Teflon sealed bottles. Aqueous calibration standards: 1 hour, unless preserved, sealed and stored at 4 °C.	Stock standards: 1 month. Store at 4 °C in Teflon sealed vials away from light. Aqueous calibration standards: 1 day.	Stock standards: 1 month. Store between -10 °C and -20 °C in Teflon sealed bottles away from light. Aqueous calibration standards: 1 hour; 24 hours if stored in sealed vials with zero headspace.
Initial Calibration	Minimum of 5 levels, lowest near but above MDL. If %RSD < 20, linearity assumed and average RF used. Alternatively use a calibration curve.	3, 4, or 5 levels (depending on calibration range), lowest 2-10 times MDL. If %RSD <10, linearity assumed and average RF used. Alternatively use a calibration curve or single point calibration. Single point calibration standard must produce response within ± 20% of unknowns.	Minimum of 3 levels, lowest near but above MDL. If %RSD < 10, assume linearity and average RF used. Alternatively use a calibration curve.	Minimum of 5 levels. If %RSD < 10, linearity assumed and average RF used. Alternatively use a calibration curve or single point calibration. Single point calibration standard must produce response within ± 20% of unknowns.

Determination of Aromatic Volatile Organic Compounds by GC (Continued)

Method ⇒ Parameter ⇓	SW-846 Method 8020	EPA 500 Series Method 503.1	EPA 600 Series Method 602	Standard Methods Method 6220
Continuing Calibration	Mid-level calibration standard run every 10 samples. If not within ± 15% of predicted response, recalibrate. Standard RT must fall within daily RT window or system is out of control. Samples injected after criteria was exceeded must be reanalyzed.	Mid-level calibration standard analyzed at the beginning of each 8 hour shift. Response must be within ± 20% of initial calibration.	QC check standard analyzed each working day. Compare results to Table 2 (see page A-51).	Analyze one or more calibration standards daily. If not within ± 20% of predicted response, recalibrate.
Surrogate Standards	Add surrogates (e.g., α,α,α-Tifluorotoluene) to encompass range of temperature program. Results must fall within laboratory established control limits.	Not specified.	Add surrogates to encompass range of temperature program. (α,α,α-Trifluorotoluene is recommended). % Recovery limits not specified	α,α,α-Trifluorotoluene % Recovery limits not specified.
Internal Standards	Optional. If used, α,α,α-Trifluorotoluene is recommended.	Optional. If used, α,α,α-Trifluorotoluene is recommended. IS area in samples must be within ± 3 times SD of calibration.	Optional. If used, α,α,α-Trifluorotoluene is recommended.	Optional. If used, α,α,α-Trifluorotoluene is recommended.
Accuracy/ Precision	One MS/MSD per 20 samples or each batch of samples, whichever is more frequent. Compare results to Table 3 (see page A-93). (See also: **QC Check Standard/ Samples.**)	No MS required. At least quarterly, analyze replicates of QC standards to determine precision. (See also: **QC Check Standard/Samples.**)	One MS per 10 samples from each site or 1 per month, whichever is more frequent. Compare %R to Table 2 (see page A-51). (See also: **QC Check Standard/ Samples.**)	One MS per 10 samples analyzed from each site or one per month, whichever is more frequent. Compare %R to Table 6220:II (see page A-73). (See also: **QC Check Standard/ Samples.**)
Blanks	One method blank per extraction batch (up to 20 samples) or when there is a change in reagents, whichever is more frequent.	One method blank per batch of samples processed as a group within a work shift. One Field Reagent Blank (FRB) per batch of samples.	One method blank per day.	One method blank per day.
Preservation/ Storage Conditions	pH ≤ 2 with HCl or H_2SO_4. If residual chlorine, add sodium thiosulfate (aqueous). Store at 4 °C.	pH ≤ 2 with HCl. If residual chlorine, add ascorbic acid or sodium thiosulfate. Store at 4 °C. Collect all samples in duplicate.	pH ≤ 2 with HCl. If residual chlorine, add sodium thiosulfate. Store at 4 °C.	pH ≤ 2 with HCl. Add ascorbic acid, sodium thiosulfate or sodium sulfite if residual chlorine. Store at 4 °C.
Holding Time (4)	14 days	14 days	14 days	14 days
Field Sample Amount Required (5)	40 mL VOA vial in duplicate without headspace (air bubbles) (aqueous) 4 oz. (solid) glass container Teflon lined septa	40 mL VOA vial in duplicate without headspace (air bubbles) glass container Teflon lined septa	40 mL VOA vial in duplicate without headspace (air bubbles) glass container Teflon lined septa	40 mL VOA vial in duplicate without headspace (air bubbles) glass container Teflon lined septa
Amount for Extraction	5 mL (aqueous) 5 grams (solid)	5 mL	5 mL	5 mL

Determination of Aromatic Volatile Organic Compounds by GC (Continued)

Method ⟹ Parameter ⟱	SW-846 Method 8020	EPA 500 Series Method 503.1	EPA 600 Series Method 602	Standard Methods Method 6220
Other Criteria (Method Specific)	When doubt exists in compound identification, second column or GC/MS confirmation should be used.	Second column confirmation is recommended. Retention times must vary by <10% over 8 hour period.	When analyzing unfamiliar samples, support identifications by at least one additional qualitative technique, such as second column confirmation or GC/MS	When analyzing unfamiliar samples, support identifications by at least one additional qualitative technique, such as second column confirmation or GC/MS

Notes:

(1) Analyte lists may vary among methods; a smaller list in one method is not necessarily a subset of a larger list in another method.

(2) Initial, one-time demonstration of ability to generate acceptable accuracy and precision. Procedure may need to be repeated if changes in instrumentation or methodology occur.

(3) Indicates maximum usage time. If comparisons to QC check standards indicate a problem, more frequent preparation may be necessary.

(4) Unless otherwise indicated, holding times are from the date of sample collection.

(5) Approximate volumes to be gathered for analysis. Additional volumes are required for the generation of QC data.

(6) MDL determination for Standard Methods 6000 Series: Analyze a minimum of seven check samples (concentration = 0.2 times MCL or 10 times estimated MDL.). Average percent recovery should be 80% to 120% of true value with %RSD ≤ 35%. Use results to determine MDLs. Broader acceptance ranges exist for some compounds with lower extraction efficiency and are indicated in the specific method.

Determination of Aromatic Volatile Organic Compounds by GC

Chemical Name	CAS Number	SW-846 Method 8020	EPA 500 Method 503.1	EPA 600 Method 602	Std. Methods Method 6220
Benzene	71-43-2	•	•	•	•
Bromobenzene	108-86-1		•		
n-Butylbenzene	104-51-8		•		
sec-Butylbenzene	135-98-8		•		
tert-Butylbenzene	98-06-6		•		
Chlorobenzene	108-90-7	•	•	•	•
2-Chlorotoluene	95-49-8		•		
4-Chlorotoluene	106-43-4		•		
1,2-Dichlorobenzene	95-50-1	•	•	•	•
1,3-Dichlorobenzene	541-73-1	•	•	•	•
1,4-Dichlorobenzene	106-46-7	•	•	•	•
Ethyl benzene	100-41-4	•	•	•	•
Hexachlorobutadiene	87-68-3		•		
Isopropylbenzene	98-82-8		•		
4-Isopropyltoluene	99-87-6		•		
Naphthalene	91-20-3		•		
n-Propylbenzene	103-65-1		•		
Styrene	100-42-5		•		
Tetrachloroethene	127-18-4		•		
Toluene	108-88-3	•	•	•	•
1,2,3-Trichlorobenzene	87-61-6		•		
1,2,4-Trichlorobenzene	120-82-1		•		
Trichloroethene	79-01-6		•		
1,2,4-Trimethylbenzene	95-63-6		•		
1,3,5-Trimethylbenzene	108-67-8		•		
m-Xylene	108-38-3		•		
o-Xylene	95-47-6		•		
p-Xylene	106-42-3		•		
Xylene (total)	1330-20-7	•			

Determination of Halogenated Volatile Organic Compounds By GC

Method ⇒ Parameter ⇓	SW-846 Method 8010A	SW-846 Method 8021	EPA 500 Series Method 502.1	EPA 500 Series Method 502.2	EPA 600 Series Method 601	Standard Methods Method 6230
Applicability	Groundwater, soils, sludge, water miscible liquid waste and non-water miscible waste	Groundwater, soils, sludge, water miscible liquid waste and non-water miscible waste & others.	Drinking water, raw source water	Drinking water, raw source water	Municipal and industrial discharge water	Most environmental water samples
Number of Analytes (1)	34 total	60 total (same as method 502.2)	40 total	60 total (same as method 8021)	29 total	28 total
Method Validation (2)	Analyze 4 replicates of QC check standard. Compare accuracy and precision results to Table 3 (see page A-91).	Analyze 4 replicates of QC check standard (parameters of interest at 20 µg/L). Compare accuracy and precision results to Table 2 (see page A-96).	Analyze 4-7 replicates of QC check sample. (Laboratory Fortified Blank) %R = 80-120. %RSD < 20. Use results to calculate MDLs. MDLs must be sufficient to detect analytes at regulatory levels.	Analyze 4-7 replicates of QC check sample. (Laboratory Fortified Blank) %R = 80-120. %RSD < 20. Use results to calculate MDLs. MDLs must be sufficient to detect analytes at regulatory levels.	Analyze 4 replicates of QC check standard. Compare accuracy and precision results to Table 2 (see page A-49).	Analyze 4 replicates of QC check standard. Compare accuracy and precision results to Table 6230:II (see page A-75) or optionally Table 6230:III (see page A-76). Determine MDL (6)
QC Check Standards/ Samples	If MS results fall outside the ranges designated in Table 3 (see page A-91), a QC check standard must be analyzed and fall within those ranges.	If MS results fall outside the ranges designated in Table 2 (see page A-96), a QC check standard must be analyzed and fall within those ranges.	Analyze one QC check sample (Laboratory Fortified Blank) per 20 samples or each batch of samples processed together within a work shift, whichever is more frequent. %R = 80-120. %RSD < 20 Use results to estimate MDLs. MDLs must be sufficient to detect analytes at regulatory levels. Analyze replicate QC check samples and a QC sample from an external source at least quarterly.	Analyze one QC check sample (Laboratory Fortified Blank) per 20 samples or each batch of samples processed together within a work shift, whichever is more frequent. %R = 80-120. %RSD < 20 Use results to estimate MDLs. MDLs must be sufficient to detect analytes at regulatory levels. Analyze replicate QC check samples and a QC sample from an external source at least quarterly.	If MS results fall outside the ranges designated in Table 2 (see page A-49), a QC check standard must be analyzed and fall within those ranges.	If MS results fall outside the ranges designated in Table 6230:II (see page A-75), a QC check standard must be analyzed and fall within those ranges.
Method Detection Limit	MDLs listed in Table 1 (see page A-89). EQLs listed in Table 2 (see page A-90).	MDLs listed in Table 1 (see page A-94). EQLs listed in Table 3 (see page A-98).	MDLs listed in Table 2 (see page A-18).	MDLs listed in Table 2 (see page A-19) and Table 4 (see page A-21).	MDLs listed in Table 1 (see page A-48).	MDLs listed in Table 6230:I (see page A-74)

Determination of Halogenated Volatile Organic Compounds By GC (Continued)

Method ⇒ / Parameter ⇓	SW-846 Method 8010A	SW-846 Method 8021	EPA 500 Series Method 502.1	EPA 500 Series Method 502.2	EPA 600 Series Method 601	Standard Methods Method 6230
Standard Solution Expiration (3)	Stock standards (except gases and reactive compounds, i.e.: 2-chloroethyl-vinyl ether): 6 months. Store between -10 °C to -20 °C with minimum headspace in Teflon sealed bottles away from light.. Stock gas and reactive compound standards: 2 months. Same conditions as stock standards. Aqueous calibration standards: 1 hour; 24 hours if stored with zero headspace.	Stock standards (except gases and reactive compounds, i.e.: 2-chloroethyl-vinyl ether): 6 months. Store between -10 °C to -20 °C with minimum headspace in Teflon sealed bottles away from light. Stock gas and reactive compound standards: 2 months. Same conditions as stock standards. Aqueous calibration standards: 1 hour; 24 hours if stored with zero headspace.	Stock and primary dilution standards (except gases): 4 weeks if stored at 4 °C in PTFE sealed bottles. Stock and primary dilution gaseous: 1 week. Store at < 0 °C; 1 day at room temperature. Calibration standards: 1 hour unless stored at 4 °C with zero headspace.	Stock and primary dilution standards (except gases): 4 weeks if stored at 4 °C in PTFE sealed bottles. Stock and primary dilution gaseous: 1 week. Store at < 0 °C; 1 day at room temperature. Calibration standards: 1 hour unless stored at 4 °C with zero headspace.	Stock standards, (except gases and reactive compounds, i.e.: 2-chloroethyl-vinyl ether): 1 month. Store between -10 °C to -20 °C with minimum headspace in Teflon sealed bottles away from light. Stock gas and reactive compound standards: 1 week. Same conditions as stock standards. Aqueous calibration standards: 1 hour; 24 hours if stored with zero headspace.	Stock standards, (except gases and 2-chloroethyl-vinyl ether): 1 month. Store between -10 °C and -20 °C in Teflon sealed bottles away from light. Stock gas and 2-chloroethylvinyl ether standards: 1 week. Store between -10 °C and -20 °C away from light. Aqueous calibration standards: 1 hour; 24 hours if stored in sealed vials with zero headspace.
Initial Calibration	Minimum of 5 levels, lowest near but above MDL. If %RSD < 20, linearity assumed and average RF used. Alternatively use a calibration curve.	Minimum of 5 levels, lowest near but above MDL. If %RSD < 20, linearity assumed and average RF used. Alternatively use a calibration curve.	3, 4, or 5 levels (depending on calibration range), lowest 2-10 times MDL. If %RSD < 10, linearity assumed and average RF used. Alternatively use a calibration curve or a single point calibration. Single point calibration standard must produce response within ± 20% of unknowns.	3, 4, or 5 levels (depending on calibration range), lowest 2-10 times MDL. If %RSD < 10, linearity assumed and average RF used. Alternatively use a calibration curve or a single point calibration. Single point calibration standard must produce response within ± 20% of unknowns.	Minimum of 3 levels, lowest near but above MDL. If %RSD < 10, linearity assumed and average RF used. Alternatively use a calibration curve.	Minimum of 3 levels, lowest near but above MDL. If %RSD < 10, linearity assumed and average RF used. Alternatively use a calibration curve.
Continuing Calibration	Mid-level calibration standard run every 10 samples. If not within ± 15% of predicted response, recalibrate. Standard RT must fall within daily RT window or system is out of control. Samples injected after criteria exceeded must be reanalyzed.	Mid-level calibration standard run every 10 samples. If not within ± 15% of predicted response, recalibrate. Standard RT must fall within daily RT window or system is out of control. Samples injected after criteria exceeded must be reanalyzed.	Analyze one or more calibration standards daily. If not within ± 20% of predicted response, recalibrate.	Analyze one or more calibration standards daily or with every batch of 20 samples, whichever is more frequent. If not within ± 20% of predicted response, recalibrate.	QC check standard analyzed each working day. Compare results to Table 2 (see page A-49).	QC check standard analyzed each working day. Compare results to Table 6230:II (see page A-75).

GENIUM PUBLISHING CORPORATION

Determination of Halogenated Volatile Organic Compounds By GC (Continued)

Method ⇒ Parameter ⇓	SW-846 Method 8010A	SW-846 Method 8021	EPA 500 Series Method 502.1	EPA 500 Series Method 502.2	EPA 600 Series Method 601	Standard Methods Method 6230
Surrogate Standards	Bromochloromethane, 2-Bromo-1-chloropropane, and 1,4-Dichlorobutane. Results must fall within lab established control limits.	At least two surrogate compounds. Recommended: Bromochloromethane, 2-Bromo-1-chloropropane, and 1,4-Dichlorobutane. Results must fall within lab established control limits.	Not specified.	Not specified.	Bromochloromethane, 2-bromo-1-chloropropane, and 1,4-dichlorobutane	Bromochloromethane, 2-bromo-1-chloropropane, and 1,4-dichlorobutane
Internal Standards	Optional. If used, exact compounds not specified, though same as used for surrogates may be appropriate.	Optional. If used, fluorobenzene and 2-bromo-1-chloropropane. Area must be ± 3 SD of calibration standards.	Optional. If used, 1,4-Dichlorobutane or 2-Bromo-1-chloropropane. Area must be ± 3 SD of calibration standards.	Optional. 1-chloro-2-fluorobenzene, 2-bromo-1-chloropropane and fluorobenzene recommended. Area must be ± 3 SD of calibration standards.	Optional. If used, surrogate compounds have been successfully used as internal standards.	Optional. 2-bromo-1-chloropropane or 1,4-dichlorobutane is recommended. Select one or more of the compounds with similar analytical behavior to compounds of interest.
Accuracy/ Precision	One MS/MSD per 20 samples or each batch of samples, whichever is more frequent. Compare results to Table 3 (see page A-91). (See also: **QC Check Standards/ Samples.**)	One MS/MSD per 20 samples or each batch of samples, whichever is more frequent. Compare results to Table 2 (see page A-96). (See also: **QC Check Standards/ Samples.**)	MS not required. Laboratory fortified blanks (LFB) should be run at least quarterly to determine precision. (See also: **QC Check Standards/ Samples.**)	MS not required. Laboratory fortified blanks (LFB) should be run at least quarterly to determine precision. (See also: **QC Check Standards/ Samples.**)	One MS per 10 samples from each site or 1 per month, whichever is more frequent. Compare %R to Table 2 (see page A-49).	One MS per 10 samples from each site or 1 per month, whichever is more frequent. Compare %R to Table 6230:II (see page A-75). (See also: **QC Check Standards/ Samples.**)
Blanks	One method blank per extraction batch (up to 20 samples) or when there is a change in reagents, whichever is more frequent.	One method blank per extraction batch (up to 20 samples) or when there is a change in reagents, whichever is more frequent.	One method blank per batch of samples processed within a workshift. Field Reagent Blank (FRB) with each sample set.	One method blank per batch of samples processed at the same time. Field Reagent Blank (FRB) with each sample set.	One method blank per day.	One method blank per day.
Preservation/ Storage Conditions	Sodium thiosulfate if residual chlorine (aqueous). Store at 4 °C.	Sodium thiosulfate if residual chlorine (aqueous). Store at 4 °C.	pH ≤ 2 with HCl. If residual chlorine, add ascorbic acid or sodium thiosulfate. Store at 4 °C. Collect all samples in duplicate.	pH ≤ 2 with HCl. Add ascorbic acid or sodium thiosulfate if residual chlorine is present. Store at 4 °C.	Sodium thiosulfate if residual chlorine. Store at 4 °C.	pH ≤ 2 with HCl. Add ascorbic acid, sodium thiosulfate or sodium sulfite if residual chlorine is present. Store at 4 °C.
Holding Time (4)	14 days	14 days	14 days	14 days	14 days	14 days

Determination of Halogenated Volatile Organic Compounds By GC (Continued)

Method ⇒ / Parameter ⇓	SW-846 Method 8010A	SW-846 Method 8021	EPA 500 Series Method 502.1	EPA 500 Series Method 502.2	EPA 600 Series Method 601	Standard Methods Method 6230
Field Sample Amount Required (5)	40 mL VOA vial in duplicate without head-space (air bubbles) (aqueous) 4 oz. (solid) glass container Teflon lined top	40 mL VOA vial in duplicate without head-space (air bubbles) (aqueous) 4 oz. (solid) glass container Teflon lined top	40 mL VOA vial in duplicate without head-space (air bubbles) glass container Teflon lined top	40 mL VOA vial in duplicate without head-space (air bubbles) glass container Teflon lined top	40 mL VOA vial in duplicate without head-space (air bubbles) glass container Teflon lined top	40 mL VOA vial in duplicate with-out headspace (air bubbles) glass container Teflon lined top
Amount for Extraction	5 mL (aqueous) 5 grams (solid)	5 mL (aqueous) 5 grams (solid)	5 mL	5 mL	5 mL	5 mL
Other Criteria (Method Specific)	When doubt exists in compound identification, second column or GC/MS confirmation should be used.	When doubt exists in compound identification, second column or GC/MS confirmation should be used.	When doubt exists in compound identification, second column confirmation or GC/MS confirmation recommended. Retention times must vary by <10% over 8 hour period.	When doubt exists in compound identification, second column or GC/MS confirmation recommended.	When doubt exists in compound identification, second column confirmation recommended.	When analyzing unfamiliar samples, support identifications by at least one additional qualitative technique.

Notes:
(1) Analyte lists may vary among methods; a smaller list in one method is not necessarily a subset of a larger list in another method.
(2) Initial, one-time demonstration of ability to generate acceptable accuracy and precision. Procedure may need to be repeated if changes in instrumentation or methodology occur.
(3) Indicates maximum usage time. If comparisons to QC check standards indicate a problem, more frequent preparation may be necessary.
(4) Unless otherwise indicted, holding times are from the date of sample collection.
(5) Approximate volumes to be gathered for analysis. Additional volumes are required for the generation of QC data.
(6) MDL determination for Standard Methods 6000 Series: Analyze a minimum of seven check samples (concentration = 0.2 times MCL or 10 times estimated MDL.). Average percent recovery should be 80% to 120% of true value with %RSD ≤ 35%. Use results to determine MDLs. Broader acceptance ranges exist for some compounds with lower extraction efficiency and are indicated in the specific method.

Determination of Halogenated Volatile Organic Compounds By GC

Chemical Name	CAS Number	SW-846 Method 8010A	SW-846 Method 8021	EPA 500 Method 502.1	EPA 500 Method 502.2	EPA 600 Method 601	Std Methods Method 6230
Benzene	71-43-2		•		•		
Benzyl chloride (α-chlorotoluene)	100-44-7	•					
Bromobenzene	108-86-1	•	•	•	•		
Bromochloromethane	74-97-5		•	•	•		
Bromodichloromethane	75-27-4	•	•	•	•	•	•
Bromoform	75-25-2	•	•	•	•	•	•
Bromomethane	74-83-9	•	•	•	•	•	•
n-Butylbenzene	104-51-8		•		•		
sec-Butylbenzene	135-98-8		•		•		
tert-Butylbenzene	98-06-6		•		•		
Carbon tetrachloride	56-23-5	•	•	•	•	•	•
Chlorobenzene	108-90-7	•	•	•	•	•	•
Chloroethane	75-00-3	•	•	•	•	•	•
2-Chloroethylvinyl ether	110-75-8	•				•	•
Chloroform	67-66-3	•	•	•	•	•	•
Chloromethane	74-87-3	•	•	•	•	•	•
2-Chlorotoluene	95-49-8		•	•	•		
4-Chlorotoluene	106-43-4		•	•	•		
1,2-Dibromo-3-chloropropane	96-12-8		•		•		
Dibromochloromethane	124-48-1	•	•	•	•	•	•
1,2-Dibromoethane	106-93-4		•	•	•		
Dibromomethane	74-95-3	•	•	•	•		
1,2-Dichlorobenzene	95-50-1	•	•	•	•	•	•
1,3-Dichlorobenzene	541-73-1	•	•	•	•	•	•
1,4-Dichlorobenzene	106-46-7	•	•	•	•	•	•
Dichlorodifluoromethane	75-71-8	•	•	•	•	•	
1,1-Dichloroethane	75-34-3	•	•	•	•	•	•
1,2-Dichloroethane	107-06-2	•	•	•	•	•	•
1,1-Dichloroethene	75-35-4	•	•	•	•	•	•
cis-1,2-Dichloroethene	156-59-2		•	•	•		
trans-1,2-Dichloroethene	156-60-5	•	•	•	•	•	•
1,2-Dichloropropane	78-87-5	•	•	•	•	•	•
1,3-Dichloropropane	142-28-9		•	•	•		
2,2-Dichloropropane	590-20-7		•	•	•		
1,1-Dichloropropene	563-58-6		•	•	•		

Determination of Halogenated Volatile Organic Compounds By GC (Continued)

Chemical Name	CAS Number	SW-846 Method 8010A	SW-846 Method 8021	EPA 500 Method 502.1	EPA 500 Method 502.2	EPA 600 Method 601	Std Methods Method 6230
cis-1,3-Dichloropropene	10061-01-5	•	•	•	•	•	•
trans-1,3-Dichloropropene	10061-02-6	•	•	•	•	•	•
Ethyl benzene	100-41-4		•		•		
Hexachlorobutadiene	87-68-3		•		•		
Isopropylbenzene	98-82-8		•		•		
4-Isopropyltoluene	99-87-6		•		•		
Methylene chloride	75-09-2	•	•	•	•	•	•
Naphthalene	91-20-3		•		•		
n-Propylbenzene	103-65-1		•		•		
Styrene	100-42-5		•		•		
1,1,1,2-Tetrachloroethane	630-20-6	•	•	•	•		
1,1,2,2-Tetrachloroethane	79-34-5	•	•	•	•	•	•
Tetrachloroethene	127-18-4	•	•	•	•	•	•
Toluene	108-88-3		•		•		
1,2,3-Trichlorobenzene	87-61-6		•		•		
1,2,4-Trichlorobenzene	120-82-1		•		•		
1,1,1-Trichloroethane	71-55-6	•	•	•	•	•	•
1,1,2-Trichloroethane	79-00-5	•	•	•	•	•	•
Trichloroethene	79-01-6	•	•	•	•	•	•
Trichlorofluoromethane	75-69-4	•	•	•	•	•	•
1,2,3-Trichloropropane	96-18-4	•	•	•	•		
1,2,4-Trimethylbenzene	95-63-6		•		•		
1,3,5-Trimethylbenzene	108-67-8		•		•		
Vinyl chloride	75-01-4	•	•	•	•	•	•
m-Xylene	108-38-3		•		•		
o-Xylene	95-47-6		•		•		
p-Xylene	106-42-3		•		•		

Analytical Methods - Inorganic Constituents

The numerous analytical methods for inorganic constituent testing have been organized into five major categories:

- Cyanide (total and amenable)
- Organic carbon (total)
- Mercury identified by cold vapor atomic absorption
- Trace metals identified by flame and graphite furnace atomic absorption spectroscopy
- Trace metals identified by inductively coupled plasma technique

Within each category there is a section called *Method Comparison* that summarizes the key method requirements and components for each testing method. The methods are compared on a side-by-side basis, which allows the user to quickly compare key aspects of the test methods. Where appropriate the user is referred to a table in Appendix A that provides details about an aspect of the test method. The tables in Appendix A are taken directly from the regulations.

For determining if a specific inorganic analyte is analyzed by a test method, each category that identifies more than one specific analyte has an *Analyte Listing* of all analytes addressed by the test methods within the category. This is also presented on a side-by-side basis so that the user can quickly see which method(s) cover the specific inorganic analyte in question.

If you want to find a test method for a specific inorganic analyte but don't know what category(ies) it belongs to, refer to Appendix C. Appendix C is an alphabetical listing of all inorganic analytes specified in the regulations/methods included in this book. Use this appendix to identify the category to turn to for additional information.

Determination of Total and Amenable Cyanide

Method ⇒ Parameter ⇓	SW-846 Methods 9010A, & 9012	EPA Series Methods 335.1, 335.2	Standard Methods Methods 4500-CN C, D, E and G	CLP CN (Inorganic SOW) Method 335.2 CLP-M
Applicability	Wastes and leachates	Drinking, surface, saline water, domestic and industrial wastes.	Wastewater, ground-waters, and solid waste.	Water, soil, and sediments
Number of Analytes	Total CN and CN amenable to chlorination	Total CN and CN amenable to chlorination	Total CN and CN amenable to chlorination	Total CN only
Method Validation (1)	Not specified.	Not specified.	Not specified.	Not specified.
QC Check Standards/ Samples	9010A: Analyze check standards with each sample batch. %R = 85-115 9012: Verify calibration with an independent check standard every 15 samples. %R = 85-115	Not specified.	Not specified.	Verify initial calibration with a distilled independent standard. %R = 85-115%.
Method Detection Limit	0.02 mg/L	0.02 mg/L	0.02 mg/L	CRDL: 0.01 mg/L
Standard Solution Expiration (2)	Stock standard: Not specified Calibration standards: Prepare daily	Stock standard: Not specified Calibration standards: Prepare daily	Stock cyanide: Check weekly Calibration standards: Prepare daily	Stock cyanide: Not specified Calibration standards: Prepare daily
Initial Calibration	Colorimetric: 6 levels plus blank	Colorimetric: 6 levels plus blank	Colorimetric: Series of standards over range 0.2 – 6 µg CN plus blank	Colorimetric: Minimum of 3 levels plus a blank (one standard at CRDL).
Continuing Calibration	9010A: Not specified 9012: Verify curve with mid-level standard every sample batch.	Not specified	Colorimetric: Recheck calibration curve period-ically, and each time new reagent used. Periodically is not de-fined in the method.	Colorimetric: Calibration checked after every 10 samples or 2 hours, whichever is more frequent. %R = 85–115%.
Surrogate Standards	Not applicable.	Not applicable.	Not applicable.	Not applicable.
Internal Standards	Not applicable.	Not applicable.	Not applicable.	Not applicable.
Accuracy/ Precision	9010A: One MS per 20 samples. %R not specified. One duplicate per 20 samples %RSD < 20. 9012: One MS/MSD per 10 samples. %R, %RSD not specified.	MS to check sample distillation efficiency; no frequency specified.	Not specified.	One MS and 1 duplicate for each sample batch or case, whichever is more frequent. %R = 75–125%. RPD < 20%
Blanks	Titration: One reagent blank per sample batch. Colorimetric: One calibration blank per calibration and one method blank per 20 samples or sample batch, whichever is more frequent.	Titration: One reagent blank per sample batch. Colorimetric: Calibration blank No distillation blank specified.	Titration: One reagent blank per sample batch. Colorimetric: Calibration blank No distillation blank specified.	Titration: One reagent blank per sample batch or case, whichever is more frequent. Colorimetric: Calibration blanks after every 10 samples or 2 hours, whichever is more frequent. Analyze a distillation blank with each sample batch.

Determination of Total and Amenable Cyanide (Continued)

Method ⇒ Parameter ⇓	SW-846 Methods 9010A, & 9012	EPA Series Methods 335.1, 335.2	Standard Methods Methods 4500-CN C, D, E and G	CLP CN (Inorganic SOW) Method 335.2 CLP-M
Preservation/ Storage Conditions	Aqueous: pH ≥ 12 with NaOH. Sodium arsenite or ascorbic acid if oxidizing agents present. Store at 4 °C (aqueous). Store at 4 °C (solid).	pH ≥ 12 with NaOH. Ascorbic acid if residual chlorine present. Store at 4 °C.	Preserve with sodium arsenite or < 0.1 g/L sodium thiosulphate if oxidizing agents present. Preserve with lead acetate or lead carbonate if sulfide ion present. Preserve with NaOH if aldehydes present. Adjust pH = 12 to 12.5 and store in cool dark place if samples not immediately analyzed.	Aqueous: pH ≥ 12 with NaOH. Ascorbic acid if residual chlorine present. Store at 2-6 °C (aqueous). Store at 2-6 °C (solid).
Holding Times (3)	14 days	14 days (24 hours when sulfide present)	Not specified.	12 days from sample receipt
Field Sample Amount Required (4)	1 liter (aqueous) 200 grams (solid) glass or polyethylene container	1 liter glass or plastic container	1 liter glass or polyethylene container	1 liter glass or polyethylene container
Amount for Extraction	500 mL (1000 mL if both total and amenable CN) (aqueous) 1-5 grams (2-10 grams if both total and amenable CN) (solid)	500 mL (1000 mL if both total and amenable CN)	500 mL (1000 mL if both total and amenable CN)	500 mL (aqueous) 1-5 grams (solid)
Other Criteria (Method Specific)	Distill a high and a low standard; results should be ± 10% of undistilled concentration. If sulfides present, all standards must be distilled. If matrix interference, use MSA.	Distill a high and a low standard; results should be ± 10% of undistilled concentration. If sulfides present, all standards must be distilled.	None.	Distill one mid-level standard; results must be ± 15% of undistilled concentration.

Notes:
(1) Initial, one-time demonstration of ability to generate acceptable accuracy and precision.
(2) Indicates maximum usage time. If comparisons to QC check standards indicate a problem, more frequent preparation may be necessary.
(3) Unless otherwise indicated, holding times are from the date of sample collection.
(4) Approximate volumes to be gathered for analysis. Additional volumes are required for the generation of QC data.

Determination of Total Organic Carbon

Method ⇒ Parameter ⇓	SW-846 Method 9060A	EPA Series Method 415.1	Standard Methods Method 5310B
Applicability	Groundwater, surface and saline waters, domestic and industrial wastes	Groundwater, surface and saline waters, domestic and industrial wastes	Ground and surface waters, wastewater
Number of Analytes	No specific compounds; includes natural sugars, mercaptans, alkanes, low molecular weight alcohols and oils, cellulose fibers, oily matter adsorbed on silt.	No specific compounds; includes natural sugars, mercaptans, alkanes, low molecular weight alcohols and oils, cellulose fibers, oily matter adsorbed on silt	No specific compounds; classes of organic compounds not specified.
Method Validation (1)	Not specified.	Not specified.	Not specified.
QC Check Standards/ Samples	Verify calibration with an independently prepared check standard. No criteria specified.	Not specified.	Not specified
Method Detection Limit	1 mg/L	1 mg/L	1 mg/L
Standard Solution Expiration (2)	Not specified.	Not specified.	Not specified.
Initial Calibration	Per instrument manufacturer's specifications.	Per instrument manufacturer's specifications.	Per instrument manufacturer's specifications.
Continuing Calibration	Not specified.	Not specified.	Not specified.
Surrogate Standards	Not applicable.	Not applicable.	Not applicable.
Internal Standards	Not applicable.	Not applicable.	Not applicable.
Accuracy/ Precision	One MS/MSD per 20 samples or per sample batch, whichever is more frequent. Analyze samples in quadruplicate.	Not specified.	Not specified.
Blanks	One calibration blank per calibration and one method blank per sample batch.	Method blanks analyzed; frequency not specified.	Method and instrument blanks analyzed; frequency not specified.
Preservation/ Storage Conditions	$pH \leq 2$ with HCl or H_2SO_4. Protect from light & atmospheric O_2. Store at 4 °C.	$pH \leq 2$ with HCl or H_2SO_4. Protect from light & atmospheric O_2. Store at 4 °C.	$pH \leq 2$ with H_2SO_4 or H_3PO_4 only if inorganic carbon is later purged. Protect from light. Store at 4 °C.
Holding Time (3)	28 days	28 days	28 days
Field Sample Amount Required (4)	Not specified. glass or polyethylene container	25 mL glass or polyethylene container	50 mL glass or polyethylene container
Amount for Extraction	Not specified.	Not specified.	Not specified.
Other Criteria (Method Specific)	None.	None.	None.

Notes:
(1) Initial, one-time demonstration of ability to generate acceptable accuracy and precision.
(2) Indicates maximum usage time. If comparisons to QC check standards indicate a problem, more frequent preparation may be necessary.
(3) Unless otherwise indicated, holding times are from the date of sample collection.
(4) Approximate volumes to be gathered for analysis. Additional volumes are required for generation of QC data.

Determination of Mercury by Cold Vapor Atomic Absorption Spectroscopy

Method ⟹ Parameter ⇩	SW-846 Methods 7470/7471	EPA Series Methods 245.1/245.5	Standard Methods Methods 3112B	CLP (Inorganic SOW)
Applicability	Drinking, surface, saline waters, domestic and industrial wastes, extracts, soils, sludges, sediments, industrial and other solid wastes	Drinking, surface, and saline waters, domestic and industrial wastes, soils, sediments, and sludges	Drinking, surface, saline, groundwaters, and wastewaters	Water, soil, and sediment
Number of Analytes	Hg only	Hg only	Hg only	Hg only
Method Validation (1)	Not specified.	Not specified.	Determine MDL	Not specified.
QC Check Standards/ Samples	Verify each calibration with an independent check standard. %R = 90-110. Analyze one laboratory control sample with each batch. (See also: **Initial Calibration**)	Performance Sample: One (blind) sample analyzed per year for metals being determined; results within EPA control limits. OPTIONAL: Known reference standards analyzed once per quarter for metals being determined.	Verify calibration with independent check standard. %R = 95-105.	Initial calibration verified with independent standard. %R = 80-120. Digested QC check standard (Laboratory Control Sample) analyzed with each sample batch or case, whichever is more frequent. %R = 80-120.
Method Detection Limit	0.0002 mg/L	0.0002 mg/L	Not specified.	CRDL: 0.0002 mg/L
Standard Solution Expiration (2)	Stock standards: Not specified. Calibration standards: Prepare fresh at time of analysis.	Stock standards: Not specified. Calibration standards: Prepare fresh at time of analysis.	Stock standards: Not specified. Calibration standards: Prepare daily.	Stock standards: Not specified. Calibration standards: Prepare fresh at time of analysis.
Initial Calibration	At least three levels plus a blank	Five levels plus a blank	Three levels plus a blank	Minimum of 4 levels plus a blank. If instrument design prevents 4 level calibration, analyze additional levels immediately after calibration. %R = 95-105 Calibrate at least daily or every 24 hours, or each time instrument is set up.
Continuing Calibration	Analyze a mid-level calibration standard or QC check standard after every 15 samples. %R = 80-120. Recalibrate after each hour of continuous analysis.	Analyze a low-level standard daily or every 20 samples, whichever is more frequent. %R = 90-110.	Analyze a mid-level standard after every 10 samples, or beginning and end of run, whichever is more frequent. %R = 90-110	Analyze mid-level independent standard after each 10 samples or every 2 hours, whichever is more frequent. %R = 80-120.
Surrogate Standards	Not applicable.	Not applicable.	Not applicable.	Not applicable.
Internal Standards	Not applicable.	Not applicable.	Not applicable.	Not applicable.

Determination of Mercury by Cold Vapor Atomic Absorption Spectroscopy (Continued)

Method ⇒ Parameter ⇓	SW-846 Methods 7470/7471	EPA Series Methods 245.1/245.5	Standard Methods Methods 3112B	CLP (Inorganic SOW)
Accuracy/ Precision	One MS/MSD per 10 samples or each batch of samples, whichever is more frequent. Analyze one diluted sample per batch. Results for diluted sample must fall within 10% of undiluted sample or MS must be run. MS %R = 85-115 or MSA must be run.	OPTIONAL: One duplicate sample per every 10 samples or per each batch of samples (if batch contains < 10). Results within EPA control limits.	Analyze one sample spiked prior to digestion and one duplicate sample per batch. %R = 85-115 Analyze one sample spiked prior to digestion per 10 samples or per analysis batch, which-ever is more frequent. %R = 85-115. Analyze one duplicate sample per analysis batch.	One MS and one duplicate per sample delivery group or per similar matrix type. %R = 75-125. %RPD < 20.
Blanks	One method blank per batch of samples processed at the same time. One calibration blank with each calibration.	Method blank not specified. Calibration blank with each calibration.	One calibration blank every 10 samples or beginning and end of run, whichever is more frequent.	One method blank per sample delivery group or per sample process batch, whichever is more frequent. Analyze calibration blanks after initial and continuing calibration verification or every 10 samples or 2 hours, whichever is more frequent. Also at the start and end of sample analysis.
Preservation/ Storage Conditions	Aqueous: pH ≤ 2 with HNO_3 Solid: Store at 4 °C.	Aqueous: pH ≤ 2 with HNO_3 Solid: Store at 4 °C.	pH ≤ 2 with HNO_3 Store at 4 °C.	Aqueous: pH ≤ 2 with HNO_3 Solid: Store at 2-6 °C.
Holding Time (3)	28 days	28 days	5 weeks	26 days from sample receipt
Field Sample Amount Required (4)	1 liter (aqueous) 200 grams (solid) glass or polyethylene container	1 liter (aqueous) 200 grams (solid) glass or polyethylene container	1 liter glass or polyethylene container	1 liter (aqueous) 200 grams (solid) glass or polyethylene container
Amount for Extraction	100 mL (aqueous) 2 grams (solids)	100 mL (aqueous) 2 grams (solids)	100 mL	100 mL (aqueous) 2 grams (solids)
Other Criteria (Method-Specific)	Use MSA to compensate for matrix interferences.	Use MSA to compensate for matrix interferences.	Use MSA to compensate for matrix interferences	CRDL standard at beginning of run. To verify calibration linearity near CRDL, analyze a CRA standard at CRDL or IDL, whichever is greater at the beginning of each run.

Notes:
(1) Initial, one-time demonstration of ability to generate acceptable accuracy and precision.
(2) Indicates maximum usage time. If comparisons to QC check standards indicate a problem, more frequent preparation may be necessary.
(3) Unless otherwise indicated, holding times are from the date of sample collection.
(4) Approximate volumes to be gathered for analysis. Additional volumes are required for the generation of QC data.

Determination of Trace Metals by Flame and Graphite Furnace Atomic Absorption Spectroscopy

Method ⇒ Parameter ⇓	SW-846 Method 7000 Series	EPA Series Method 200 Series	Standard Methods Method 3000 Series	CLP (Inorganic SOW)
Applicability	Drinking, surface, saline waters, domestic and industrial wastes, extracts, soils, sludges, sediments, industrial and other solid wastes	Drinking, surface, and saline waters; domestic and industrial wastes	Drinking, surface, saline, groundwaters, and wastewaters	Water, soil, and sediment
Number of Analytes (1)	27 total	33 total	39 total	22 total (including alternate methods)
Method Validation (2)	Not specified.	Not specified.	Determine MDL	Not specified.
QC Check Standards/ Samples	Verify calibration with an independent check standard. %R = 90-110. Laboratory control sample processed with each sample batch.	Performance Sample: One (blind) sample analyzed per year for metals being determined; results within EPA control limits. OPTIONAL: Known reference standards analyzed once per quarter for metals being determined	Verify calibration with independent check standard. %R = 95-105.	Initial calibration verified with independent standard. %R = 90-110. Digested QC check standard (Laboratory Control Sample) analyzed with each sample batch or case, whichever is more frequent. %R = 80-120
Method Detection Limit	Detection limits listed in Table 1 of method 7000A (see page A-88).	Detection limits listed in Table 1 of Metals Introductory chapter (see page A-16).	FLAA: IDLs listed in Table 3111:I of Method 3111 (see page A-65). GFAA: IDLs listed in Table 3113:II of Method 3113 (see page A-66).	CRDLs listed in Exhibit C (see page A-15). IDLs determined quarterly.
Standard Solution Expiration (3)	Stock standards: Not specified. Calibration standards: Prepare fresh at time of analysis.	Stock standards: Not specified. Calibration standards: Prepare fresh at time of analysis.	Stock standards: Not specified. Calibration standards: Prepare daily.	Stock standards: Not specified. Calibration standards: Not specified.
Initial Calibration	Minimum of 3 levels plus a blank.	Minimum of 3 levels plus a blank.	Minimum of 3 levels plus a blank.	Minimum of 4 levels plus a blank. If instrument design prevents 4 level calibration, analyze additional levels immediately after calibration. %R = 95-105 Calibrate at least daily or every 24 hours, or each time instrument is set up.
Continuing Calibration	Analyze a mid-level calibration standard or QC check standard after every 10 samples. %R = 80-120 or reanalyze previous 10 samples.	Analyze a low-level standard daily or every 20 samples, whichever is more frequent. %R = 90-110.	Analyze a mid-level standard after every 10 samples, or beginning and end of each run, whichever is more frequent. %R = 90-110	Analyze mid-level independent standard after each 10 samples or every 2 hours, whichever is more frequent. %R = 90-110.
Surrogate Standards	Not applicable.	Not applicable.	Not applicable.	Not applicable.
Internal Standards	Not applicable.	Not applicable.	Not applicable.	Not applicable.

Determination of Trace Metals by Flame and Graphite Furnace Atomic Absorption Spectroscopy (Continued)

Method ⇒ Parameter ⇓	SW-846 Method 7000 Series	EPA Series Method 200 Series	Standard Methods Method 3000 Series	CLP (Inorganic SOW)
Accuracy/ Precision	One MS/MSD per batch of samples processed at the same time. Criteria not specified	OPTIONAL: One duplicate sample per every 10 samples, or per each set of samples (if set contains < 10). Results within EPA control limits.	Analyze one sample spiked prior to digestion per 10 samples or analysis batch, whichever is more frequent. %R = 85-115 Analyze one duplicate sample per analysis batch. GFAA only: Analyze all digestates in duplicate.	One MS and one duplicate per sample delivery group or per similar matrix type. %R = 75-125. %RPD < 20.
Blanks	One method blank per batch of samples processed at the same time. One calibration blank with curve.	Method blank not specified. Calibration blank with each calibration.	One method preparation blank per batch of samples processed at the same time. One calibration blank every 10 samples or beginning and end of run, whichever is more frequent.	One method blank per sample delivery group or per sample process batch, whichever is more frequent. Analyze calibration blanks after initial and continuing calibration verification or every 10 samples or 2 hours, whichever is more frequent. Also at start and end of sample analysis.
Preservation/ Storage Conditions	Aqueous: pH ≤ 2 with HNO$_3$ Solid: Store at 4 °C.	pH ≤ 2 with HNO$_3$	pH ≤ 2 with HNO$_3$	Aqueous: pH ≤ 2 with HNO$_3$ Solid: Store at 2-6 °C.
Holding Time (4)	6 months	6 months	6 months	180 days from sample receipt
Field Sample Amount Required (5)	1 liter (aqueous) 200 grams (solid) glass or polyethylene container	1 liter (aqueous) 200 grams (solid) glass or polyethylene container	1 liter glass or polyethylene container	1 liter (aqueous) 200 grams (solid) glass or polyethylene container
Amount for Extraction	100 mL (aqueous) 2 grams (solid)	100 mL (aqueous) 2 grams (solid)	100 mL	100 mL (aqueous) 1 gram (solid) Microwave digestion: 45 mL (aqueous) 0.5 grams (solid)
Other Criteria (Method Specific)	Test for matrix interference with each matrix using serial dilution (if >25 times detection limit). Results must agree between undiluted ±10%, or post-digestion spike must agree. %R = 85%-115%. Use standard additions to compensate for matrix interferences.	For GFAA, verify the absence of interferences by diluting and spiking each matrix. Use standard additions to compensate for matrix interferences.	Test for matrix interference with each matrix by analyzing undiluted and diluted samples.	CRDL standard at beginning of run. To verify calibration linearity near CRDL analyze a CRA Standard at CRDL or IDL, whichever is greater, at the beginning of each run. All GFAA digestates analyzed in duplicate. %RPD < 20. All GFAA digestates spiked to determine matrix interferences. If verified, MSA used.

Notes:
(1) Analyte lists may vary among methods; a smaller list in one method is not necessarily a subset of a larger list in another method.
(2) Initial, one-time demonstration of ability to generate acceptable accuracy and precision.
(3) Indicates maximum usage time. If comparisons to QC check standards indicate a problem, more frequent preparation may be necessary.
(4) Unless otherwise indicated, holding times are from the date of sample collection.
(5) Approximate volumes to be gathered for analysis. Additional volumes are required for the generation of QC data.

Trace Metals by Flame and Graphite Furnace Atomic Absorption Spectroscopy

Metal	Analytical Method Type[1]	SW-846 Method 7000 Series	EPA Method 200 Series	Std. Methods Method 3000	CLP Inorganic Method[2]
Aluminum	AAS	7020	202.1	3111	
Aluminum	GFAA		202.2	3113	202.2
Antimony	AAS	7040	204.1	3111	204.1
Antimony	GFAA	7041	204.2	3113	204.2
Arsenic	GFAA	7060	206.2	3113	206.2
Arsenic	GHAA	7061A	206.3	3114	
Barium	AAS	7080	208.1	3111	208.1
Barium	GFAA	7081	208.2	3113	208.2
Beryllium	AAS	7090	210.1	3111	210.1
Beryllium	GFAA	7091	210.2	3113	210.2
Bismuth	AAS			3111	
Cadmium	AAS	7130	213.1	3111	213.1
Cadmium	GFAA	7131	213.2	3113	213.2
Calcium	AAS	7140	215.1	3111	215.1
Cesium	AAS			3111	
Chromium	AAS	7190	218.1	3111	218.1
Chromium	GFAA	7191	218.2	3113	218.2
Cobalt	AAS	7200	219.1	3111	219.1
Cobalt	GFAA	7201	219.2	3113	219.2
Copper	AAS	7210	220.1	3111	220.1
Copper	GFAA	7211	220.2	3113	220.2
Gold	AAS		231.1	3111	
Gold	GFAA		231.2		
Iridium	AAS		235.1	3111	
Iridium	GFAA		235.2		
Iron	AAS	7380	236.1	3111	236.1
Iron	GFAA	7381	236.2	3113	236.2
Lead	AAS	7420	239.1	3111	239.1
Lead	GFAA	7421	239.2	3113	239.2
Lithium	AAS	7430		3111	
Magnesium	AAS	7450	242.1	3111	242.1
Manganese	AAS	7460	243.1	3111	243.1
Manganese	GFAA	7461	243.2	3113	243.2
Molybdenum	AAS	7480	246.1	3111	
Molybdenum	GFAA	7481	246.2	3113	
Nickel	AAS	7520	249.1	3111	249.1
Nickel	GFAA		249.2	3113	249.2
Osmium	AAS	7550	252.1	3111	
Osmium	GFAA		252.2		

Trace Metals by Flame and Graphite Furnace Atomic Absorption Spectroscopy (Continued)

Metal	Analytical Method Type[1]	SW-846 Method 7000 Series	EPA Method 200 Series	Std. Methods Method 3000	CLP Inorganic Method[2]
Palladium	AAS		253.1	3111	
Palladium	GFAA		253.2		
Platinum	AAS		255.1	3111	
Platinum	GFAA		255.2		
Potassium	AAS	7610	258.1	3111	258.1
Rhenium	AAS		264.1	3111	
Rhenium	GFAA		264.2		
Rhodium	AAS		265.1	3111	
Rhodium	GFAA		265.2		
Ruthenium	AAS		267.1	3111	
Ruthenium	GFAA		267.2		
Selenium	GFAA	7740	270.2	3113	270.2
Selenium	GHAA	7741	270.3	3114	
Silicon/Silica (SiO$_2$)	AAS			3111	
Silver	AAS	7760A	272.1	3111	272.1
Silver	GFAA	7761	272.2	3113	272.2
Sodium	AAS	7770	273.1	3111	273.1
Sodium	GFAA		273.2		
Strontium	AAS	7780		3111	
Thallium	AAS	7840	279.1	3111	279.1
Thallium	GFAA	7841	279.2		279.2
Thorium	AAS			3111	
Tin	AAS	7870	282.1	3111	
Tin	GFAA		282.2	3113	
Titanium	AAS		283.1	3111	
Titanium	GFAA		283.2		
Vanadium	AAS	7910	286.1	3111	286.1
Vanadium	GFAA	7911	286.2		286.2
Zinc	AAS	7950	289.1	3111	289.1
Zinc	GFAA	7951	289.2		289.2

(1) Analytical Method Type: AAS -   Flame Atomic Absorption Spectroscopy (also called FLAA)
                                     GFAA -   Graphite Furnace Atomic Absorption Spectroscopy
                                     GHAA -   Gaseous Hydride Atomic Absorption Spectroscopy

(2) All methods in this column have the suffix "CLP-M", indicating the method has been modified for the Contract Laboratory Program

Determination of Trace Metals by ICP

Method ⇒ Parameter ⇓	SW-846 Method 6010A	EPA Series Method 200.7	Standard Methods Method 3120	CLP (Inorganic SOW) Method 200.7 CLP-M
Applicability	Nearly all matrices, including groundwater, extracts, industrial wastes, soils, sludges, sediments and solid wastes	Drinking water, surface, saline and industrial and domestic wastes	Water and wastewater	Water, soil, and sediment
Number of Analytes (1)	26 total	25 total	27 total	22 total
Method Validation (2)	Not specified.	Not specified.	Not specified.	Not specified. Interelement correction analysis must be run at start of contract and annually thereafter, or after instrument adjustment.
QC Check Standards/ Samples	Verify calibration with an independently pre-pared check standard.	Verify each calibration with QC check standard from external source. If recovery is not 95% - 105%, recalibrate.	Analyze QC check standard with each sample batch. %R = 95-105 of expected values.	Initial calibration verified with independent standard. %R = 90-110. Digested QC check standard analyzed with each sample batch or case, whichever is more frequent. %R = 80-120.
Method Detection Limit	IDLs listed in Table 1 (see page A-87). MDLs vary by matrix.	IDLs listed in Table 1 (see page A-17). MDLs vary by matrix.	IDLs listed in Table 3120:I (see page A-67).	CRDLs listed in Exhibit C (see page A-15). IDLs determined quarterly.
Standard Solution Expiration (3)	Not specified. Initially verify with QC standard and monitor weekly thereafter.	Not specified.	Not specified.	Not specified. Initially verify with QC standard and monitor weekly thereafter.
Initial Calibration	Per instrument manu-facturer's specifications (should consist of 3 levels and a blank).	Per instrument manu-facturer's specifications.	Per instrument manufacturer's specifications.	Per instrument manufacturer's specifications.
Continuing Calibration	Analyze mid-level cali-bration standard after each 10 samples and at the end of the analytical run. %R = 90-110.	Analyze mid-level cali-bration standard after each 10 samples. %R = 95-105.	Analyze 2 mg/L instrument check stan-dard after each 10 samples. %R = 95-105.	Analyze mid-level standard after each 10 samples or every 2 hours, whichever is more frequent. %R = 90-110.
Surrogate Standards	Not applicable.	Not applicable.	Not applicable.	Not applicable.
Internal Standards	Not applicable.	Not applicable.	Not applicable.	Not applicable.
Accuracy/ Precision	One MS/MSD per 20 samples or each batch of samples, whichever is more frequent. %R = 80-120. %RPD < 20 Duplicate sample after each 20 samples or analytical batch, whichever is more frequent.	Not specified.	Analyze one method QC check sample with each run. %R = 95-105	One MS and one duplicate per sample delivery group or per similar matrix type. %R = 75-125. %RPD < 20.

Determination of Trace Metals by ICP (Continued)

Method ⇒ Parameter ⇓	SW-846 Method 6010A	EPA Series Method 200.7	Standard Methods Method 3120	CLP (Inorganic SOW) Method 200.7 CLP-M
Blanks	One method blank per batch of samples processed at the same time. Analyze calibration blank after initial and continuing calibration verification or each 10 samples.	One method blank per batch of samples processed at the same time. Analyze calibration blank after each 10 samples.	Method blank analyzed; frequency not specified. Analyze one calibration blank between each sample or standard.	One method blank per sample delivery group or per sample process batch, whichever is more frequent. Analyze calibration blanks after initial and continuing calibration verification or every 10 samples or 2 hours, whichever is more frequent. Also at start and end of sample analysis.
Preservation/ Storage Conditions	Aqueous: pH $\leq$ 2 with HNO_3 Solid: Store at 4 °C.	pH $\leq$ 2 with HNO_3	pH $\leq$ 2 with HNO_3	Aqueous: pH $\leq$ 2 with HNO_3 Solid: Store at 4 °C.
Holding Time (4)	6 months	6 months	6 months	180 days from sample receipt.
Field Sample Amount Required (5)	1 liter (aqueous) 200 grams (solid) glass or polyethylene container	1 liter (aqueous) 200 grams (solid) glass or polyethylene container	1 liter glass or polyethylene container	1 liter (aqueous) 200 grams (solid) glass or polyethylene container
Amount for Extraction	100 mL (aqueous) 2 grams (solid)	100 mL (aqueous) 2 grams (solid)	100 mL	100 mL (aqueous) 1 grams (solid) Microwave digestion: 45 mL (aqueous) 0.5 grams (solid)
Other Criteria (Method Specific)	Analyze ICS (Interference Check Sample) at beginning and end of run or twice during 8 hour shift, whichever is more frequent. Results ± 20% of true value. Reanalyze highest standard after calibration. Results ± 5% of true value. Test each new matrix for interference: Perform 1:4 dilution on sample containing analytes >10 times IDL. Results of dilution should agree within ±10% of original measurement. Perform post - digestion spike at 10 to 100 times IDL. %R = 75-125	Analyze ICS (Interference Check Sample) at beginning, end, and periodically during run. Results within 1.5 times SD of mean.	Determine interference correction factors each time samples are analyzed. Test for matrix interference with each matrix using serial dilution (> 1 mg/L) or post-digestion spike (<1 mg/L).	Analyze ICS at beginning and end of run or twice during 8 hour shift, whichever is more frequent. Results ± 20% of true value. To verify linearity near the CRDL, CRI standard at 2 times CRDL or 2 times IDL, whichever is greater, run at beginning and end of run, or minimum of twice per 8 hours (no criteria). Serial dilution analysis must be performed on a sample from each matrix. If concentration > 50 times IDL, must agree within ±10% of original. Verify linear ranges quarterly by running Linear Range Analysis (LRA) standard. Results must be within 5% of true value. All measurements minimum of 2 replicate exposures, report average.

Notes:
(1) Analyte lists may vary among methods: a smaller list in one method is not necessarily a subset of a larger list in another method.
(2) Initial, one-time demonstration of ability to generate acceptable accuracy and precision.
(3) Indicates maximum usage time. If comparisons to QC check standard indicate a problem, more frequent preparation may be necessary.
(4) Unless otherwise indicated, holding times are from the date of sample collection.
(5) Approximate volumes to be gathered for analysis. Additional volumes are required for the generation of QC data.

Determination of Trace Metals by ICP

Chemical Name	SW-846 Method 6010A	EPA Series Method 200.7	Std. Methods Method 3120	CLP (Inorganic SOW) Method 200.7 CLP-M
Aluminum	•	•	•	•
Antimony	•	•	•	•
Arsenic	•	•	•	•
Barium	•	•	•	•
Beryllium	•	•	•	•
Boron		•	•	
Cadmium	•	•	•	•
Calcium	•	•	•	•
Chromium	•	•	•	•
Cobalt	•	•	•	•
Copper	•	•	•	•
Iron	•	•	•	•
Lead	•	•	•	•
Lithium	•		•	
Magnesium	•	•	•	•
Manganese	•	•	•	•
Molybdenum	•	•	•	
Nickel	•	•	•	•
Phosphorus	•			
Potassium	•	•	•	•
Selenium	•	•	•	•
Silica (SiO_2)		•	•	
Silver	•	•	•	•
Sodium	•	•	•	•
Strontium	•		•	
Thallium	•	•	•	•
Vanadium	•	•	•	•
Zinc	•	•	•	•

APPENDICES

APPENDICES

APPENDIX A

These are a compilation of some of the tables, exhibits, etc., found in the methods. If the method comparison refers to a specific table, that table will be found in this appendix. Only tables and exhibits referred to in the method comparison are found in this appendix.

Listing of Tables and Exhibits from the Methods Included in this Appendix

APPENDIX A (Continued)

APPENDIX A (Continued)

APPENDIX A (Continued)

Pesticides – Organic Statement of Work – *Exhibit C*

Target Compound List (TCL) and Contract Required Quantitation Limits (CRQL)

Pesticides/Aroclors	CAS Number	Quantitation Limits*		
		Water µg/L	Soil µg/Kg	On Column (pg)
α-BHC	319-84-6	0.05	1.7	5
β-BHC	319-85-7	0.05	1.7	5
δ-BHC	319-86-8	0.05	1.7	5
γ-BHC (Lindane)	58-89-9	0.05	1.7	5
Heptachlor	76-44-8	0.05	1.7	5
Aldrin	309-00-2	0.05	1.7	5
Heptachlor epoxide	1024-57-3	0.05	1.7	5
Endosulfan I	959-98-8	0.05	1.7	5
Dieldrin	60-57-1	0.10	3.3	10
4,4'-DDE	72-55-9	0.10	3.3	10
Endrin	72-20-8	0.10	3.3	10
Endosulfan II	33213-65-9	0.10	3.3	10
4,4'-DDD	72-54-8	0.10	3.3	10
Endosulfan sulfate	1031-07-8	0.10	3.3	10
4,4'-DDT	50-29-3	0.10	3.3	10
Methoxychlor	72-43-5	0.50	17.0	50
Endrin ketone	53494-70-5	0.10	3.3	10
Endrin aldehyde	7421-36-3	0.10	3.3	10
α-Chlordane	5103-71-9	0.05	1.7	5
γ-Chlordane	5103-74-2	0.05	1.7	5
Toxaphene	8001-35-2	5.0	170.0	500
Aroclor-1016	12674-11-2	1.0	33.0	100
Aroclor-1221	11104-28-2	2.0	67.0	200
Aroclor-1232	11141-16-5	1.0	33.0	100
Aroclor-1242	53469-21-9	1.0	33.0	100
Aroclor-1248	12672-29-6	1.0	33.0	100
Aroclor-1254	11097-69-1	1.0	33.0	100
Aroclor-1260	11096-82-5	1.0	33.0	100

*Quantitation limits listed for soil/sediment are based on wet weight. The quantitation limits calculated by the laboratory for soil/sediment, calculated on dry weight basis as required by the contract, will be higher.

There is no differentiation between the preparation of low and medium soil samples in this method for the analysis of Pesticides/Aroclors.

Matrix Spike Recovery and Relative Percent Difference Limits

Compound	%Recovery Water	RPD Water	%Recovery Soil	RPD Soil
γ-BHC (Lindane)	56-123	15	46-127	50
Heptachlor	40-131	20	35-130	31
Aldrin	40-120	22	34-132	43
Dieldrin	52-126	18	31-134	38
Endrin	56-121	21	42-139	45
4,4'-DDT	38-127	27	23-134	50

Target Compound List (TCL) and Contract Required Quantitation Limits (CRQL)

Semivolatiles	CAS Number	Quantitation Limits*			
		Water µg/L	Low Soil µg/Kg	Med. Soil µg/Kg	On Column (ng)
Phenol	108-95-2	10	330	10000	(20)
bis(2-Chloroethyl) ether	111-44-4	10	330	10000	(20)
2-Chlorophenol	95-57-8	10	330	10000	(20)
1,3-Dichlorobenzene	541-73-1	10	330	10000	(20)
1,4-Dichlorobenzene	106-46-7	10	330	10000	(20)
1,2-Dichlorobenzene	95-50-1	10	330	10000	(20)
2-Methylphenol	95-48-7	10	330	10000	(20)
2,2'-oxybis (1-Chloropropane)#	108-60-1	10	330	10000	(20)
4-Methylphenol	106-44-5	10	330	10000	(20)
N-Nitroso-di-n-propylamine	621-64-7	10	330	10000	(20)
Hexachloroethane	67-72-1	10	330	10000	(20)
Nitrobenzene	98-95-3	10	330	10000	(20)
Isophorone	78-59-1	10	330	10000	(20)
2-Nitrophenol	88-75-5	10	330	10000	(20)
2,4-Dimethylphenol	105-67-9	10	330	10000	(20)
bis(2-Chloroethoxy) methane	111-91-1	10	330	10000	(20)
2,4-Dichlorophenol	120-83-2	10	330	10000	(20)
1,2,4-Trichlorobenzene	120-82-1	10	330	10000	(20)
Naphthalene	91-20-3	10	330	10000	(20)
4-Chloroaniline	106-47-8	10	330	10000	(20)
Hexachlorobutadiene	87-68-3	10	330	10000	(20)
4-Chloro-3-methylphenol	59-50-7	10	330	10000	(20)
2-Methylnaphthalene	91-57-6	10	330	10000	(20)
Hexachlorocyclopentadiene	77-47-4	10	330	10000	(20)
2,4,6-Trichlorophenol	88-06-2	10	330	10000	(20)
2,4,5-Trichlorophenol	95-95-4	25	800	25000	(50)
2-Chloronaphthalene	91-58-7	10	330	10000	(20)
2-Nitroaniline	88-74-4	25	800	25000	(50)
Dimethyl phthalate	131-11-3	10	330	10000	(20)
Acenaphthylene	208-96-8	10	330	10000	(20)
2,6-Dinitrotoluene	606-20-2	10	330	10000	(20)
3-Nitroaniline	99-09-2	25	800	25000	(50)
Acenaphthene	83-32-9	10	330	10000	(20)

Semi-volatile Organic Analysis – Organic Statement of Work – *Exhibit C*

Target Compound List (TCL) and Contract Required Quantitation Limits (CRQL) (Continued)

Semivolatiles	CAS Number	Quantitation Limits*			
		Water µg/L	Low Soil µg/Kg	Med. Soil µg/Kg	On Column (ng)
2,4-Dinitrophenol	51-28-5	25	800	25000	(50)
4-Nitrophenol	100-02-7	25	800	25000	(50)
Dibenzofuran	132-64-9	10	330	10000	(20)
2,4-Dinitrotoluene	121-14-2	10	330	10000	(20)
Diethylphthalate	84-66-2	10	330	10000	(20)
4-Chlorophenyl-phenyl ether	7005-72-3	10	330	10000	(20)
Fluorene	86-73-7	10	330	10000	(20)
4-Nitroaniline	100-01-6	25	800	25000	(50)
4,6-Dinitro-2-methylphenol	534-52-1	25	800	25000	(50)
N-nitrosodiphenylamine	86-30-6	10	330	10000	(20)
4-Bromophenyl-phenylether	101-55-3	10	330	10000	(20)
Hexachlorobenzene	118-74-1	10	330	10000	(20)
Pentachlorophenol	87-86-5	25	800	25000	(50)
Phenanthrene	85-01-8	10	330	10000	(20)
Anthracene	120-12-7	10	330	10000	(20)
Carbazole	86-74-8	10	330	10000	(20)
Di-n-butylphthalate	84-74-2	10	330	10000	(20)
Fluoranthene	206-44-0	10	330	10000	(20)
Pyrene	129-00-0	10	330	10000	(20)
Butylbenzylphthalate	85-68-7	10	330	10000	(20)
3,3'-Dichlorobenzidine	91-94-1	10	330	10000	(20)
Benzo(a)anthracene	56-55-3	10	330	10000	(20)
Chrysene	218-01-9	10	330	10000	(20)
bis(2-Ethylhexyl)phthalate	117-81-7	10	330	10000	(20)
Di-n-octylphthalate	117-84-0	10	330	10000	(20)
Benzo(b)fluoranthene	205-99-2	10	330	10000	(20)
Benzo(k)fluoranthene	207-08-9	10	330	10000	(20)
Benzo(a)pyrene	50-32-8	10	330	10000	(20)
Indeno(1,2,3-cd)pyrene	193-39-5	10	330	10000	(20)
Dibenz(a,h)anthracene	53-70-3	10	330	10000	(20)
Benzo(g,h,i)perylene	191-24-2	10	330	10000	(20)

Previously known by the name bis(2-Chloroisopropyl) ether
*Quantitation limits listed for soil/sediment are based on wet weight. The quantitation limits calculated by the laboratory for soil/sediment, calculated on dry weight basis as required by the contract, will be higher.

Semi-volatile Organic Analysis – Organic Statement of Work – *Table 1*

DFTPP Key Ions and Ion Abundance Criteria for Quadrapole Mass Spectrometers

Mass	Ion Abundance Criteria
51	30.0-80.0 percent of mass 198
68	Less than 2.0 percent of mass 69
69	Present
70	Less than 2.0 percent of mass 69
127	25.0-75.0 percent of mass 198
197	Less than 1.0 percent of mass 198
198	Base Peak, 100 percent relative abundance (see note)
199	5.0-9.0 percent of mass 198
275	10.0-30.0 percent of mass 198
365	Greater than 0.75 percent of mass 198
441	Present but less than mass 443
442	40.0-110.0 percent of mass 198
443	15.0-24.0 percent of mass 442

NOTE: All ion abundances MUST be normalized to m/z 198, the nominal base peak, even though the ion abundances of m/z 442 may be up to 110 percent that of m/z 198.

Semi-volatile Organic Analysis – Organic Statement of Work – *Table 5*

Relative Response Factor Criteria for Initial and Continuing Calibration of Semivolatile Target Compounds

Semivolatile Compounds	Minimum RRF	Maximum %RSD	Maximum %Diff
Phenol	0.800	20.5	25.0
bis(2-Chloroethyl)ether	0.700	20.5	25.0
2-Chlorophenol	0.800	20.5	25.0
1,3-Dichlorobenzene	0.600	20.5	25.0
1,4-Dichlorobenzene	0.500	20.5	25.0
1,2-Dichlorobenzene	0.400	20.5	25.0
2-Methylphenol	0.700	20.5	25.0
4-Methylphenol	0.600	20.5	25.0
N-Nitroso-di-n-propylamine	0.500	20.5	25.0
Hexachloroethane	0.300	20.5	25.0
Nitrobenzene	0.200	20.5	25.0
Isophorone	0.400	20.5	25.0
2-Nitrophenol	0.100	20.5	25.0
2,4-Dimethylphenol	0.200	20.5	25.0
bis(2-Chloroethoxy)methane	0.300	20.5	25.0
2,4-Dichlorophenol	0.200	20.5	25.0
1,2,4-Trichlorobenzene	0.200	20.5	25.0
Naphthalene	0.700	20.5	25.0
4-Chloro-3-methylphenol	0.200	20.5	25.0
2-Methylnaphthalene	0.400	20.5	25.0
2,4,6-Trichlorophenol	0.200	20.5	25.0
2,4,5-Trichlorophenol	0.200	20.5	25.0
2-Chloronaphthalene	0.800	20.5	25.0
Acenaphthylene	1.300	20.5	25.0
2,6-Dinitrotoluene	0.200	20.5	25.0
Acenaphthene	0.800	20.5	25.0
Dibenzofuran	0.800	20.5	25.0
2,4-Dinitrotoluene	0.200	20.5	25.0
4-Chlorophenyl-phenylether	0.400	20.5	25.0
Fluorene	0.900	20.5	25.0
4-Bromophenyl-phenylether	0.100	20.5	25.0
Hexachlorobenzene	0.100	20.5	25.0
Pentachlorophenol	0.050	20.5	25.0

Relative Response Factor Criteria for Initial and Continuing Calibration of Semivolatile Target Compounds (Continued)

Semivolatile Compounds	Minimum RRF	Maximum %RSD	Maximum %Diff
Phenanthrene	0.700	20.5	25.0
Anthracene	0.700	20.5	25.0
Fluoranthene	0.600	20.5	25.0
Pyrene	0.600	20.5	25.0
Benzo(a)anthracene	0.800	20.5	25.0
Chrysene	0.700	20.5	25.0
Benzo(b)fluoranthene	0.700	20.5	25.0
Benzo(k)fluoranthene	0.700	20.5	25.0
Benzo(a)pyrene	0.700	20.5	25.0
Indeno(1,2,3-cd)pyrene	0.500	20.5	25.0
Dibenzo(a,h)anthracene	0.400	20.5	25.0
Benzo(g,h,i)perylene	0.500	20.5	25.0
Nitrobenzene-d_5	0.200	20.5	25.0
2-Fluorobiphenyl	0.700	20.5	25.0
Terphenyl-d_{14}	0.500	20.5	25.0
Phenol-d_5	0.800	20.5	25.0
2-Fluorophenol	0.600	20.5	25.0
2-Chlorophenol-d_4	0.800	20.5	25.0
1,2-Dichlorobenzene-d_4	0.400	20.5	25.0

Surrogate Recovery Limits

Compound	%Recovery Water	%Recovery Soil
Nitrobenzene-d_5	35-114	23-120
2-Fluorobiphenyl	43-116	30-115
Terphenyl-d_{14}	33-141	18-137
Phenol-d_5	10-110	24-113
2-Fluorophenol	21-110	25-121
2,4,6-Tribromophenol	10-123	19-122
2-Chlorophenol-d_4	33-110	20-130 (advisory)
1,2-Dichlorobenzene-d_4	16-110	20-130 (advisory)

Matrix Spike Recovery and Relative Percent Difference Limits

Compound	%Recovery Water	RPD Water	%Recovery Soil	RPD Soil
Phenol	12-110	42	26-90	35
2-Chlorophenol	27-123	40	25-102	50
1,4-Dichlorobenzene	36-97	28	28-104	27
N-Nitroso-di-n-propylamine	41-116	38	41-126	38
1,2,4-Trichlorobenzene	39-98	28	38-107	23
4-Chloro-3-methylphenol	23-97	42	26-103	33
Acenaphthene	46-118	31	31-137	19
4-Nitrophenol	10-80	50	11-114	50
2,4-Dinitrotoluene	24-96	38	28-89	47
Pentachlorophenol	9-103	50	17-109	47
Pyrene	26-127	31	35-142	36

Target Compound List (TCL) and Contract Required Quantitation Limits (CRQL)

Volatiles	CAS Number	Quantitation Limits*			
		Water µg/L	Low Soil µg/Kg	Med. Soil µg/Kg	On Column (ng)
Chloromethane	74-87-3	10	10	1200	(50)
Bromomethane	74-83-9	10	10	1200	(50)
Vinyl chloride	75-01-4	10	10	1200	(50)
Chloroethane	75-00-3	10	10	1200	(50)
Methylene chloride	75-09-2	10	10	1200	(50)
Acetone	67-64-1	10	10	1200	(50)
Carbon disulfide	75-15-0	10	10	1200	(50)
1,1-Dichloroethene	75-35-4	10	10	1200	(50)
1,1-Dichloroethane	75-34-3	10	10	1200	(50)
1,2-Dichloroethene (total)	540-59-0	10	10	1200	(50)
Chloroform	67-66-3	10	10	1200	(50)
1,2-Dichloroethane	107-06-2	10	10	1200	(50)
2-Butanone	78-93-3	10	10	1200	(50)
1,1,1-Trichloroethane	71-55-6	10	10	1200	(50)
Carbon tetrachloride	56-23-5	10	10	1200	(50)
Bromodichloromethane	75-27-4	10	10	1200	(50)
1,2-Dichloropropane	78-87-5	10	10	1200	(50)
cis-1,3-Dichloropropene	10061-01-5	10	10	1200	(50)
Trichloroethene	79-01-6	10	10	1200	(50)
Dibromochloromethane	124-48-1	10	10	1200	(50)
1,1,2-Trichloroethane	79-00-5	10	10	1200	(50)
Benzene	71-43-2	10	10	1200	(50)
trans-1,3-Dichloropropene	10061-02-6	10	10	1200	(50)
Bromoform	75-25-2	10	10	1200	(50)
4-Methyl-2-pentanone	108-10-1	10	10	1200	(50)
2-Hexanone	591-78-6	10	10	1200	(50)
Tetrachloroethene	127-18-4	10	10	1200	(50)
Toluene	108-88-3	10	10	1200	(50)
1,1,2,2-Tetrachloroethane	79-34-5	10	10	1200	(50)
Chlorobenzene	108-90-7	10	10	1200	(50)
Ethyl benzene	100-41-4	10	10	1200	(50)
Styrene	100-42-5	10	10	1200	(50)
Xylenes (total)	1330-20-7	10	10	1200	(50)

*Quantitation limits listed for soil/sediment are based on wet weight. The quantitation limits calculated by the laboratory for soil/sediment, calculated on dry weight basis as required by the contract, will be higher.

BFB Key Ions and Ion Abundance Criteria

Mass	Ion Abundance Criteria
50	8.0-40.0 percent of mass 95
75	30.0-66.0 percent of mass 95
95	Base Peak, 100 percent relative abundance
96	5.0-9.0 percent of mass 95 (see note)
173	less than 2.0 percent of mass 174
174	50.0-120.0 percent of mass 95
175	4.0-9.0 percent of mass 174
176	93.0-101.0 percent of mass 174
177	5.0-9.0 percent of mass 176

Note: All ion abundances must be normalized to m/z 95, the nominal base peak, even though the ion abundance of m/z 174 may be up to 120 percent that of m/z 95.

Relative Response Factor Criteria for Initial and Continuing Calibration of Volatile Organic Compounds

Volatile Compound	Minimum RRF	Maximum %RSD	Maximum %Diff
Bromomethane	0.100	20.5	25.0
Vinyl chloride	0.100	20.5	25.0
1,1-Dichloroethene	0.100	20.5	25.0
1,1-Dichloroethane	0.200	20.5	25.0
Chloroform	0.200	20.5	25.0
1,2-Dichloroethane	0.100	20.5	25.0
1,1,1-Trichloroethane	0.100	20.5	25.0
Carbon tetrachloride	0.100	20.5	25.0
Bromodichloromethane	0.200	20.5	25.0
cis-1,3-Dichloropropene	0.200	20.5	25.0
Trichloroethene	0.300	20.5	25.0
Dibromochloromethane	0.100	20.5	25.0
1,1,2-Trichloroethane	0.100	20.5	25.0
Benzene	0.500	20.5	25.0
trans-1,3-Dichloropropene	0.100	20.5	25.0
Bromoform	0.100	20.5	25.0
Tetrachloroethene	0.200	20.5	25.0
1,1,2,2-Tetrachloroethane	0.500	20.5	25.0
Toluene	0.400	20.5	25.0
Chlorobenzene	0.500	20.5	25.0
Ethylbenzene	0.100	20.5	25.0
Styrene	0.300	20.5	25.0
Xylenes (total)	0.300	20.5	25.0
Bromofluorobenzene	0.200	20.5	25.0

Volatile Organic Analysis – Organic Statement of Work – *Table 6*

System Monitoring Compound Recovery Limits

Compound	%Recovery Water	%Recovery Soil
Toluene-d$_8$	88-110	84-138
Bromofluorobenzene	86-115	59-113
1,2-Dichloroethane-d$_4$	76-114	70-121

Matrix Spike Recovery and Relative Percent Difference Limits

Compound	%Recovery Water	RPD Water	%Recovery Soil	RPD Soil
1,1-Dichloroethane	61-145	14	59-172	22
Trichloroethene	71-120	14	62-137	24
Benzene	76-127	11	66-142	21
Toluene	76-125	13	59-139	21
Chlorobenzene	75-130	13	60-133	21

Inorganic Target Analyte List (TAL)

Analyte	Contract Required Detection Limit (µg/L) [1,2]
Aluminum	200
Antimony	60
Arsenic	10
Barium	200
Beryllium	5
Cadmium	5
Calcium	5000
Chromium	10
Cobalt	50
Copper	25
Iron	100
Lead	3
Magnesium	5000
Manganese	15
Mercury	0.2
Nickel	40
Potassium	5000
Selenium	5
Silver	10
Sodium	5000
Thallium	10
Vanadium	50
Zinc	20
Cyanide	10

[1] Subject to the restrictions specified in the first page of Part G, Section IV of Exhibit D (Alternate Methods - Catastrophic Failure) any analytical method specified in SOW Exhibit D may be utilized as long as the documented instrument or method detection limits meet the Contract Required Detection Limit (CRDL) requirements. Higher detection limits may only be used in the following circumstance:

> If the sample concentration exceeds five times the detection limit of the instrument or method in use, the value may be reported even though the instrument or method detection limit may not equal the Contract Required Detection Limit. This is illustrated in the example below:

> **For lead:**

> Method in use = ICP Instrument Detection Limit (IDL) = 40 Sample concentration = 220 Contract Required Detection Limit (CRDL) = 3

> The value of 220 may be reported even though the instrument detection limit is greater than CRDL. The instrument or method detection limit must be documented as described in Exhibits B and E.

[2] The CRDL's are the instrument detection limits obtained in pure water that must be met using the procedure in Exhibit E. The detection limits for samples may be considerably higher depending on the sample matrix.

Method 200 Series – *Table 1*

Atomic Absorption Concentration Ranges[1]

Metal	Direct Aspiration			Furnace Procedure[4,5]	
	Detection Limit mg/L	Sensitivity mg/L	Optimum Concentration Range mg/L	Detection Limit µg/L	Optimum Concentration Range µg/L
Aluminum	0.1	1	5-50	3	20-200
Antimony	0.2	0.5	1-40	3	20-300
Arsenic[2]	0.002		0.002-0.02	1	5-100
Barium(p)	0.1	0.4	1-20	2	10-200
Beryllium	0.005	0.025	0.05-2	0.2	1-30
Cadmium	0.005	0.025	0.05-2	0.1	0.5-10
Calcium	0.01	0.08	0.2-7		
Chromium	0.05	0.25	0.5-10	1	5-100
Cobalt	0.05	0.2	0.5-5	1	5-100
Copper	0.02	0.1	0.2-5	1	5-100
Gold	0.1	0.25	0.5-20	1	5-100
Iridium(p)	3	8	20-500	30	100-1500
Iron	0.03	0.12	0.3-5	1	5-100
Lead	0.1	0.5	1-20	1	5-100
Magnesium	0.001	0.007	0.02-0.5		
Manganese	0.01	0.05	0.1-3	0.2	1-30
Mercury[3]	0.0002		0.0002-0.01		
Molybdenum(p)	0.1	0.4	1-40	1	3-60
Nickel(p)	0.04	0.15	0.3-5	1	5-50
Osmium	0.3	1	2-100	20	50-500
Palladium(p)	0.1	0.25	0.5-15	5	20-400
Platinum(p)	0.2	2	5-75	20	100-2000
Potassium	0.01	0.04	0.1-2		
Rhenium(p)	5	15	50-1000	200	500-5000
Rhodium(p)	0.05	0.3	1-30	5	20-400
Ruthenium	0.2	0.5	1-50	20	100-2000
Selenium[2]	0.002		0.002-0.02	2	5-100
Silver	0.01	0.06	0.1-4	0.2	1-25
Sodium	0.002	0.015	0.03-1		
Thallium	0.1	0.5	1-20	1	5-100
Tin	0.8	4	10-300	5	20-300
Titanium(p)	0.4	2	5-100	10	50-500
Vanadium(p)	0.2	0.8	2-100	4	10-200
Zinc	0.005	0.02	0.05-1	0.05	0.2-4

[1]   The concentrations shown are not contrived values and should be obtainable with any satisfactory atomic absorption spectrophotometer.
[2]   Gaseous hydride method.
[3]   Cold vapor technique.
[4]   For furnace sensitivity values consult instrument operating manual.
[5]   The listed furnace values are those expected when using a 20 µL injection and normal gas flow except in the case of arsenic and selenium where gas interrupt is used. The symbol (p) indicates the use of pyrolytic graphite with the furnace procedure.

Method 200.7 – *Table 1*

Recommended Wavelengths[1] and Estimated Instrumental Detection Limits

Element	Wavelength, nm	Estimated Detection Limit (μg/L)[2]
Aluminum	308.215	45
Arsenic	193.696	53
Antimony	206.833	32
Barium	455.403	2
Beryllium	313.042	0.3
Boron	249.773	5
Cadmium	226.502	4
Calcium	317.933	10
Chromium	267.716	7
Cobalt	228.616	7
Copper	324.754	6
Iron	259.940	7
Lead	220.353	42
Magnesium	279.079	30
Manganese	257.610	2
Molybdenum	202.030	8
Nickel	231.604	15
Potassium	766.491	see [3]
Selenium	196.026	75
Silica (SiO_2)	288.158	58
Silver	328.068	7
Sodium	588.995	29
Thallium	190.864	40
Vanadium	292.402	8
Zinc	213.856	2

[1] The wavelengths listed are recommended because of their sensitivity and overall acceptance. Other wavelengths may be substituted if they can provide the needed sensitivity and are treated with the same corrective techniques for spectral interference.

[2] The estimated instrumental detection limits as shown are taken from Inductively Coupled Plasma-Atomic Emission Spectroscopy-Prominent Lines,EPA-600/4-79-017. They are given as a guide for an instrumental limit. The actual method detection limits are sample dependent and may vary as the sample matrix varies.

[3] Highly dependent on operating conditions and plasma position.

Method 502.1 – *Table 2*

Single Laboratory Accuracy, Precision, and Method Detection Limits for Volatile Halogenated Organic Compounds in Water

Analyte	Concentration (µg/L)	Average Recovery (%)	Number of Samples	Rel. Std. Dev. (%)	Method Det. Limit (µg/L)
Bromobenzene	0.40	93	20	12	(a)
Bromochloromethane	0.40	90	19	9.5	(a)
Bromodichloromethane	0.20	100	17	6.5	0.003
Bromoform	0.20	95	17	15.0	0.05
Carbon tetrachloride	0.20	90	17	7.0	0.003
Chlorobenzene	0.40	88	18	9.3	0.005
Chlorocyclohexane	0.40	93	21	8.3	(a)
1-Chlorocyclohexene	0.40	93	21	12.8	(a)
Chloroethane	0.40	93	20	18	0.008
2-Chloroethylethyl ether	0.40	95	18	7.5	0.02
Chloromethane	0.40	93	16	8.5	0.01
2-Chlorotoluene	0.40	85	20	9.3	(a)
Dibromochloromethane	0.20	95	17	7.0	0.008
1,2-Dibromoethane	0.40	93	18	12.5	0.04
Dibromomethane	0.40	100	5	8.0	(a)
1,2-Dichlorobenzene	0.40	95	21	13	(a)
1,3-Dichlorobenzene	0.40	95	21	8.3	(a)
1,4-Dichlorobenzene	0.40	90	20	13	(a)
Dichlorodifluoromethane	0.40	103	12	20	(a)
1,1-Dichloroethane	0.20	95	17	6.0	0.003
1,2-Dichloroethane	0.20	110	17	7.0	0.002
1,1-Dichloroethene	0.40	88	18	9.3	0.003
1,2-Dichloroethene[b]	0.40	88	20	7.0	0.002
1,2-Dichloropropane	0.40	95	20	3.5	(a)
1,3-Dichloropropane	0.40	98	21	6.5	(a)
1,1-Dichloropropene	0.40	88	18	9.3	(a)
Methylene chloride	0.20	85	17	12.0	(a)
1,1,1,2-Tetrachloroethane	0.40	93	20	8.0	(a)
1,1,2,2-Tetrachloroethane	0.40	95	18	9.0	0.01
Tetrachloroethene	0.20	90	17	9.5	0.001
1,1,1-Trichloroethane	0.40	93	20	8.0	0.003
1,1,2-Trichloroethane	0.40	95	15	6.0	0.007
Trichloroethene	0.20	94	17	6.0	0.001
Trichlorofluoromethane	0.40	90	21	9.3	(a)
1,2,3-Trichloropropane	0.40	100	20	9.5	(a)
Vinyl chloride	0.20	110	12	15	0.01

[a]   Not determined
[b]   Includes cis- and trans- isomers

Method 502.2 – *Table 2*

Single Laboratory Accuracy, Precision and Method Detection Limits for Volatile Organic Compounds in Reagent Water for Column 1[a][b]

Analyte	Photoionization Detector			Electrolytic Conductivity Detector		
	Average Recovery, %	Rel. Std. Deviation	MDL (μg/L)	Average Recovery, %	Rel. Std. Deviation	MDL (μg/L)
Benzene	99	1.2	0.01	(c)	(c)	(c)
Bromobenzene	99	1.7	0.01	97	2.7	0.03
Bromochloromethane	(c)	(c)	(c)	96	3.0	0.01
Bromodichloromethane	(c)	(c)	(c)	97	2.9	0.02
Bromoform	(c)	(c)	(c)	106	5.2	1.6
Bromomethane	(c)	(c)	(c)	97	3.8	1.1
n-Butylbenzene	100	4.4	0.02	(c)	(c)	(c)
sec-Butylbenzene	97	2.7	0.02	(c)	(c)	(c)
tert-Butylbenzene	98	2.3	0.06	(c)	(c)	(c)
Carbon tetrachloride	(c)	(c)	(c)	92	3.6	0.01
Chlorobenzene	100	1.0	0.01	103	3.6	0.01
Chloroethane	(c)	(c)	(c)	96	3.9	0.1
Chloroform	(c)	(c)	(c)	98	2.5	0.02
Chloromethane	(c)	(c)	(c)	96	9.2	0.03
2-Chlorotoluene	ND	ND	ND	97	2.7	0.01
4-Chlorotoluene	101	1.0	0.02	97	3.2	0.01
1,2-Dibromo-3-chloropropane	(c)	(c)	(c)	86	11.3	3.0
Dibromochloromethane	(c)	(c)	(c)	102	3.3	0.3
1,2-Dibromoethane	(c)	(c)	(c)	97	2.8	0.8
Dibromomethane	(c)	(c)	(c)	109	6.7	2.2
1,2-Dichlorobenzene	102	2.1	0.05	100	1.5	0.02
1,3-Dichlorobenzene	104	1.6	0.02	106	4.0	0.02
1,4-Dichlorobenzene	103	2.1	0.01	98	2.3	0.01
Dichlorodifluoromethane	(c)	(c)	(c)	89	6.6	0.05
1,1-Dichloroethane	(c)	(c)	(c)	100	5.7	0.07
1,2-Dichloroethane	(c)	(c)	(c)	100	3.8	0.03
1,1-Dichloroethene	100	2.4	ND	103	2.8	0.07
cis-1,2-Dichloroethene	ND	ND	0.02	105	3.3	0.01
trans-1,2-Dichloroethene	93	4.0	0.05	99	3.7	0.06
1,2-Dichloropropane	(c)	(c)	(c)	103	3.7	0.01
1,3-Dichloropropane	(c)	(c)	(c)	100	3.4	0.03
2,2-Dichloropropane	(c)	(c)	(c)	105	3.4	0.05
1,1-Dichloropropene	103	3.5	0.02	103	3.3	0.02
Ethyl benzene	101	1.4	0.01	(c)	(c)	(c)
Hexachlorobutadiene	99	9.5	0.06	98	8.3	0.02

ND= Not determined.

Method 502.2 – *Table 2*

Single Laboratory Accuracy, Precision and Method Detection Limits for Volatile Organic Compounds in Reagent Water for Column 1[a][b] (Continued)

Analyte	Photoionization Detector			Electrolytic Conductivity Detector		
	Average Recovery, %	Rel. Std. Deviation	MDL (µg/L)	Average Recovery, %	Rel. Std. Deviation	MDL (µg/L)
Isopropylbenzene	98	0.9	0.05	(c)	(c)	(c)
Methylene chloride	(c)	(c)	(c)	97	2.9	0.02
Naphthalene	102	6.2	0.06	(c)	(c)	(c)
n-Propylbenzene	103	2.0	0.01	(c)	(c)	(c)
Styrene	104	1.3	0.01	(c)	(c)	(c)
1,1,1,2-Tetrachloroethane	(c)	(c)	(c)	99	2.3	0.01
1,1,2,2-Tetrachloroethane	(c)	(c)	(c)	99	6.8	0.01
Tetrachloroethene	101	1.8	0.05	97	2.5	0.04
Toluene	99	0.8	0.01	(c)	(c)	(c)
1,2,3-Trichlorobenzene	106	1.8	ND	98	3.1	0.03
1,2,4-Trichlorobenzene	104	2.2	0.02	102	2.1	0.03
1,1,1-Trichloroethane	(c)	(c)	(c)	104	3.3	0.03
1,1,2-Trichloroethane	(c)	(c)	(c)	109	5.6	ND
Trichloroethene	100	0.78	0.02	96	3.6	0.01
Trichlorofluoromethane	(c)	(c)	(c)	96	3.5	0.03
1,2,3-Trichloropropane	(c)	(c)	(c)	99	2.3	0.04
1,2,4-Trimethylbenzene	99	1.2	0.05	(c)	(c)	(c)
1,3,5-Trimethylbenzene	101	1.4	0.01	(c)	(c)	(c)
Vinyl chloride	109	5.0	0.02	95	5.9	0.04
o-Xylene	99	0.8	0.02	(c)	(c)	(c)
m-Xylene	100	1.4	0.01	(c)	(c)	(c)
p-Xylene	99	0.9	0.01	(c)	(c)	(c)

[a]   Recoveries and relative standard deviations were determined from seven samples fortified at at 10 µg/L of each analyte. Recoveries were determined by internal standard method. Internal standards were: Fluorobenzene for PID, 2-Bromo-1-chloropropane for ELCD.

[b]   Column 1: 60m long x 0.75mm ID VOCOL (Supelco, Inc.) wide-bore capillary column with 1.5 µm film thickness.

[c]   Detector does not respond.

ND= Not determined.

Method 502.2 – *Table 4*

Single Laboratory Accuracy, Precision and Method Detection Limits for Volatile Organic Compounds in Reagent Water for Column 2[a][b]

Analyte	Photoionization Detector			Electrolytic Conductivity Detector		
	Average Recovery, %	Rel. Std. Deviation	MDL (µg/L)	Average Recovery, %	Rel. Std. Deviation	MDL (µg/L)
Bromobenzene	98	1.1	0.04	96	3.2	0.14
Bromochloromethane	(c)	(c)	(c)	95	2.5	0.01
Bromodichloromethane	(c)	(c)	(c)	96	2.6	0.10
Bromoform	(c)	(c)	(c)	98	4.0	0.09
Bromomethane	(c)	(c)	(c)	97	2.4	0.19
n-Butylbenzene	95	2.4	0.03	(c)	(c)	(c)
sec-Butylbenzene	96	2.1	0.03	(c)	(c)	(c)
tert-Butylbenzene	98	2.1	0.06	(c)	(c)	(c)
Carbon tetrachloride	(c)	(c)	(c)	97	2.4	0.02
Chlorobenzene	98	1.5	0.02	98	2.2	ND
Chloroethane	(c)	(c)	(c)	97	3.2	0.13
Chloroform	(c)	(c)	(c)	92	4.2	0.01
Chloromethane	(c)	(c)	(c)	98	2.3	0.10
2-Chlorotoluene	94	3.1	0.03	99	2.3	0.04
4-Chlorotoluene	97	1.6	0.02	97	2.3	0.07
1,2-Dibromo-3-chloropropane	(c)	(c)	(c)	97	2.2	0.05
Dibromochloromethane	(c)	(c)	(c)	99.	2.0	0.05
1,2-Dibromoethane	(c)	(c)	(c)	99	2.8	0.17
Dibromomethane	(c)	(c)	(c)	98	3.5	0.10
1,2-Dichlorobenzene	97	1.4	0.03	98	2.0	0.04
1,3-Dichlorobenzene	97	1.6	0.02	97	2.2	0.07
1,4-Dichlorobenzene	97	1.5	0.03	97	2.2	0.04
Dichlorodifluoromethane	(c)	(c)	(c)	96	3.2	0.29
1,1-Dichloroethane	(c)	(c)	(c)	97	2.3	0.03
1,2-Dichloroethane	(c)	(c)	(c)	98	1.8	0.03
1,1-Dichloroethene	96	2.2	0.01	97	2.3	0.04
cis-1,2-Dichloroethene	97	1.7	0.03	96	3.3	0.05
trans-1,2-Dichloroethene	97	1.8	0.03	98	1.5	0.05
1,2-Dichloropropane	(c)	(c)	(c)	98	1.8	0.03
1,3-Dichloropropane	(c)	(c)	(c)	100	1.3	0.02
2,2-Dichloropropane	(c)	(c)	(c)	95	14.2	ND
1,1-Dichloropropene	96	2.1	0.05	97	2.6	0.02
cis-1,3-Dichloropropene	98	1.6	0.06	98	2.0	0.08
trans-1,3-Dichloropropene	99	1.7	0.06	97	1.4	0.10

ND= Not determined.

Method 502.2 – *Table 4*

Single Laboratory Accuracy, Precision and Method Detection Limits for Volatile Organic Compounds in Reagent Water for Column 2[a][b] (Continued)

Analyte	Photoionization Detector			Electrolytic Conductivity Detector		
	Average Recovery, %	Rel. Std. Deviation	MDL (μg/L)	Average Recovery, %	Rel. Std. Deviation	MDL (μg/L)
Hexachlorobutadiene	95	2.6	0.09	97	2.3	0.05
Isopropylbenzene	97	1.4	0.02	(c)	(c)	(c)
4-Isopropyltoluene	96	2.0	0.02	(c)	(c)	(c)
Methylene chloride	(c)	(c)	(c)	100	3.1	0.01
Naphthalene	96	2.1	0.02	(c)	(c)	(c)
n-Propylbenzene	97	1.8	0.03	(c)	(c)	(c)
Styrene	96	1.9	0.10	(c)	(c)	(c)
1,1,1,2-Tetrachloroethane	(c)	(c)	(c)	98	2.2	ND
1,1,2,2-Tetrachloroethane	(c)	(c)	(c)	100	2.8	0.02
Tetrachloroethene	97	1.6	0.04	97	1.9	0.02
Toluene	98	1.3	0.02	(c)	(c)	(c)
1,2,3-Trichlorobenzene	95	2.3	0.05	98	2.8	0.06
1,2,4-Trichlorobenzene	94	3.0	0.06	96	2.5	0.08
1,1,1-Trichloroethane	(c)	(c)	(c)	96	2.6	0.01
1,1,2-Trichloroethane	(c)	(c)	(c)	99	1.6	0.04
Trichloroethene	97	1.7	0.03	98	1.2	0.06
Trichlorofluoromethane	(c)	(c)	(c)	97	6.0	0.34
1,2,3-Trichloropropane	(c)	(c)	(c)	100	2.0	0.02
1,2,4-Trimethylbenzene	96	2.0	0.02	(c)	(c)	(c)
1,3,5-Trimethylbenzene	98	1.6	0.03	(c)	(c)	(c)
Vinyl chloride	95	1.1	0.01	96	2.6	0.18
o-Xylene	98	1.1	0.02	(c)	(c)	(c)
m-Xylene	98	1.1	0.02	(c)	(c)	(c)
p-Xylene	98	0.9	0.02	(c)	(c)	(c)

[a] Recoveries and relative standard deviations were determined from seven samples fortified at at 10 μg/L of each analyte. Recoveries were determined by external standard method.

[b] Column 2: 105m long x 0.53mm ID RTX-502.2 (RESTEK Corporation) mega-bore capillary column with 3.0 μm film thickness.

[c] Detector does not respond.

ND = Not determined.

Method 503.1 – *Table 2*

Single Laboratory Accuracy, Precision and Method Detection Limits for Volatile Aromatic and Unsaturated Organic Compounds in Water[a]

Analyte	Concentration Level (µg/L)	Number of Samples	Average Recovery (%)	Rel. Std. Dev. (%)	Method Det. Limit (µg/L)
Benzene	0.40	13	100	2.8	0.02
Bromobenzene	0.50	19	93	6.2	0.002
n-Butylbenzene	0.40	7	78	15.7	0.02
sec-Butylbenzene	0.40	7	80	11.0	0.02
tert-Butylbenzene	0.40	7	88	8.7	0.006
Chlorobenzene	0.50	19	96	5.8	0.004
1-Chlorocyclohexene[b]	0.50	19	89	7.1	0.008
2-Chlorotoluene					0.008
4-Chlorotoluene	0.50	17	91	5.0	
1,2-Dichlorobenzene	0.50	18	92	7.1	0.02
1,3-Dichlorobenzene	0.50	19	91	8.5	0.006
1,4-Dichlorobenzene	0.50	19	95	6.4	0.006
Ethylbenzene	0.40	7	93	8.5	0.002
Hexachlorobutadiene	0.50	10	74	16.8	0.02
Isopropylbenzene	0.40	7	88	8.7	0.005
4-Isopropyltoluene					0.009
Naphthalene	0.50	16	92	14.8	0.04
n-Propylbenzene	0.40	7	83	9.3	0.009
Styrene					0.008
Tetrachloroethene	0.50	19	97	7.8	0.01
Toluene	0.40	13	94	6.6	0.02
1,2,3-Trichlorobenzene	0.50	18	85	10.4	0.03
1,2,4-Trichlorobenzene	0.50	18	86	10.1	0.03
Trichloroethene	0.50	19	97	6.8	0.01
α,α,α,-Trifluorotoluene[c]	0.50	18	88	9.7	0.02
1,2,4-Trimethylbenzene	0.40	7	75	8.7	0.006
1,3,5-Trimethylbenzene	0.50	10	92	8.7	0.003
m-Xylene	0.40	7	90	7.7	0.004
o-Xylene	0.40	7	90	7.2	0.004
p-Xylene	0.40	7	85	8.7	0.002

[a] Matrices tested include drinking water and raw source water.
[b] Not a method analyte.
[c] Recommended internal standard.

Method 508 – *Table 2*

Single Laboratory Accuracy, Precision and Estimated Detection Limits (EDLS) for Analytes from Reagent Water and Synthetic Groundwaters[a]

Analyte	EDL[b] μg/L	Conc. μg/L	Reagent Water R[c]	Reagent Water S_R[d]	Synthetic Water 1[e] R	Synthetic Water 1[e] S_R	Synthetic Water 2[f] R	Synthetic Water 2[f] S_R
Aldrin	0.075	0.15	86	9.5	100	11.0	69	9.0
Chlordane-α	0.0015	0.15	99	11.9	96	12.5	99	7.9
Chlordane-γ	0.0015	0.15	99	11.9	96	12.5	99	6.9
Chlorneb	0.5	5	97	11.6	95	6.7	75	8.3
Chlorobenzilate	5	10	108	5.4	98	10.8	102	9.2
Chlorthalonil	0.025	0.25	91	8.2	103	10.3	71	9.2
DCPA	0.025	0.25	103	12.4	100	13.0	101	6.1
4,4'-DDD	0.0025	0.25	107	6.4	96	8.6	101	7.1
4,4'-DDE	0.01	0.1	99	11.9	96	12.5	99	6.9
4,4'-DDT	0.06	0.6	112	16.8	98	11.8	84	8.4
Dieldrin	0.02	0.2	87	8.7	103	9.3	82	7.4
Endosulfan I	0.015	0.15	87	8.7	102	8.2	84	8.4
Endosulfan sulfate	0.015	0.15	102	15.3	94	1.3	72	12.2
Endrin	0.015	0.15	88	8.8	98	9.8	104	9.4
Endrin aldehyde	0.025	0.25	88	7.9	103	11.3	84	9.2
Endosulfan II	0.024	0.15	92	10.1	98	10.8	76	6.8
Etridiazole	0.025	0.25	103	6.2	91	6.4	98	3.9
HCH-α	0.025	0.05	92	10.1	106	7.4	86	7.7
HCH-β	0.01	0.1	95	6.7	92	5.5	100	6.0
HCH-δ	0.01	0.1	102	11.2	99	11.9	103	6.2
HCH-γ	0.015	0.15	89	9.8	115	6.9	85	7.7
Heptachlor	0.01	0.1	98	11.8	85	11.1	85	7.7
Heptachlor epoxide	0.015	0.15	87	8.7	103	7.2	82	9.8
Hexachlorobenzene	0.0077	0.05	99	21.8	82	9.8	68	4.8
Methoxychlor	0.05	0.5	105	13.7	101	10.1	104	6.2
cis-Permethrin	0.5	5	91	9.1	96	11.5	86	9.5
trans-Permethrin	0.5	5	111	6.7	97	9.7	102	7.1
Propachlor	0.5	5	103	9.3	116	4.6	95	7.6
Trifluralin	0.025	0.25	103	5.2	86	10.3	87	9.6

[a] Data corrected for amount detected in blank and represent the mean of 7-8 samples.

[b] EDL = estimated detection limit; defined as either MDL (Appendix B to 40 CFR Part 136 -Definition and Procedure for the Determination of the Method Detection Limit - Revision 1.11) or a level of compound in a sample yielding a peak in the final extract with signal-to-noise ratio of approximately 5, whichever value is higher. The concentration level used in determining the EDL is not the same as the concentration level presented in this table.

[c] R = average percent recovery

[d] S_R = standard deviation of the percent recovery

[e] Corrected for amount found in blank; Absopure Nature Artesian Spring Water obtained from the Absopure Water Company in Plymouth, Michigan.

[f] Corrected for amount found in blank; reagent water fortified with fulvic acid at the 1 mg/L concentration level. A well-characterized fulvic acid, available from the International Humic Substances Society (associated with the United States Geological Survey in Denver, Colorado), was used.

Method 508 – *Table 3*

Laboratory Performance Check Solution

Test	Analyte	Conc, μg/L	Requirements
Sensitivity	Chlorpyrifos	0.0020	Detection of analyte; S/N > 3
Chromatographic Performance	DCPA	0.0500	PSF between 0.80 and 1.15[a]
Column Performance	Chlorothalonil	0.0500	Resolution > 0.50[b]
	HCH-delta	0.0400	

[a] PGF - peak Gaussian factor. Calculated using the equation: $PGF = ((1.83) \times W(1/2))/(W(1/10))$. Where $W(1/2)$ is the peak width at half height and $W(1/10)$ is the peak width at tenth height.

[b] Resolution between the two peaks as defined by the equation: $R = t/W$. Where t is the difference in elution times between the two peaks and W is the average peak width, at the baseline, of the two peaks.

Method 515.1 – *Table 2*

Single Laboratory Accuracy, Precision and Estimated Detection Limits (EDLS) for Analytes from Reagent Water and Synthetic Groundwaters[a]

Analyte	EDL µg/L [b]	Concentration µg/L	Reagent Water R[c]	Reagent Water S_R[d]	Synthetic Water[e] R	Synthetic Water[e] S_R
Acifluorfen	0.096	0.2	121	15.7	103	20.6
Bentazon	0.2	1	120	16.8	82	37.7
Chloramben	0.093	0.4	111	14.4	112	10.1
2,4-D	0.2	1	131	27.5	110	5.5
Dalapon	1.3	10	100	20	128	30.7
2,4-DB	0.8	4	87	13.1	0	0
DCPA acid metabolites	0.02	0.2	74	9.7	81	21.9
Dicamba	0.081	0.4	135	32.4	92	17.5
3,5-Dichlorobenzoic acid	0.061	0.6	102	16.3	82	7.4
Dichlorprop	0.26	2	107	20.3	106	5.3
Dinoseb	0.19	0.4	42	14.3	89	13.4
5-Hydroxydicamba	0.04	0.2	103	16.5	88	5.3
4-Nitrophenol	0.13	1	131	23.6	127	34.3
Pentachlorophenol (PCP)	0.076	0.04	130	31.2	84	9.2
Picloram	0.14	0.6	91	15.5	97	23.3
2,4,5-T	0.08	0.4	117	16.4	96	3.8
2,4,5-TP	0.075	0.2	134	30.8	105	6.3

(a)   Data corrected for amount detected in blank and represent the mean of 7-8 samples.
(b)   EDL = estimated detection limit; defined as either MDL (Appendix B to 40 CFR Part 136 - Definition and Procedure for the Determination of the Method Detection Limit - Revision 1.11) or a level of compound in a sample yielding a peak in the final extract with signal-to-noise ratio of approximately 5, whichever value is higher. The concentration used in determining the EDL is not the same as the concentration presented in this table.
(c)   R = average percent recovery
(d) S_R = standard deviation of the percent recovery
(e)   Corrected for amount found in blank; Absopure Nature Artesian Spring Water obtained from the Absopure Water Company in Plymouth, Michigan.

Method 524.1 – *Table 2*

Ion Abundance Criteria for 4-Bromofluorobenzene (BFB)

Mass (M/z)	Relative Abundance Criteria
50	15-40% of mass 95
75	30-80% of mass 95
95	Base Peak, 100% Relative Abundance
96	5-9% of mass 95
173	<2% of mass 174
174	>50% of mass 95
175	5-9% of mass 174
176	> 95% but < 101% of mass 174
177	5-9% of mass 176

Method 524.1 – *Table 3*

Accuracy and Precision Data from Seven to Nine Determinations of the Method Analytes in Reagent Water[a]

Compound	True Conc. (µg/L)	Mean Observed Conc. (µg/L)	Std. Dev. (µg/L)	Rel. Std. Dev. (%)	Mean Accuracy (% of True Value)	Method Det. Limit (µg/L)
Benzene	1.0	0.97	0.036	3.7	97	0.1
Bromobenzene	1.0	0.92	0.042	4.6	92	0.1
Bromodichloromethane	1.0	1.0	0.17	17	100	0.5
Bromoform	2.5	2.4	0.23	9.6	100	0.7
Carbon tetrachloride	1.0	0.88	0.098	11	88	0.3
Chlorobenzene	1.0	1.02	0.047	4.6	102	0.1
Chloroform	1.0	1.03	0.086	8.3	103	0.2
Dibromochloromethane	1.0	0.92	0.14	15	92	0.4
1,2-Dibromo-3-chloropropane	3.5	3.5	0.63	18	100	2
1,2-Dibromoethane	1.0	0.93	0.13	14	93	0.4
Dibromomethane	1.0	0.94	0.11	12	94	0.3
1,2-Dichlorobenzene	5.0	5.0	0.35	7.0	100	1
1,4-Dichlorobenzene	5.0	5.6	0.73	13	112	2
Dichlorodifluoromethane	1.0	0.96	0.11	12	96	0.3
1,1-Dichloroethane	1.0	1.05	0.060	5.7	105	0.2
1,2-Dichloroethane	1.0	0.97	0.077	7.9	97	0.2
1,1-Dichloroethene	1.0	1.09	0.066	6.1	109	0.2
trans-1,2-Dichloroethene	1.0	0.98	0.066	6.7	98	0.2
1,2-Dichloropropane	1.0	1.01	0.060	5.9	101	0.2
1,3-Dichloropropane	1.0	1.00	0.033	3.3	100	0.1
Methylene chloride	1.0	0.99	0.45	46	99	1
Styrene	1.0	1.2	0.072	6.0	120	0.2
1,1,2,2-Tetrachloroethane	1.0	1.11	0.14	13	111	0.4
Tetrachloroethene	1.0	0.93	0.10	11	93	0.3
Toluene	1.0	1.05	0.043	4.1	105	0.1
1,1,1-Trichloroethane	1.0	1.05	0.093	8.9	105	0.3
Trichloroethene	1.0	0.90	0.12	13	90	0.4
Trichlorofluoromethane	1.0	1.09	0.072	6.6	109	0.2
Vinyl chloride	1.0	0.98	0.11	11	98	0.3
o-Xylene	1.0	1.02	0.068	6.7	102	0.2
p-Xylene	1.0	1.11	0.047	4.2	111	0.3

[a] Data obtained by Robert W. Slater with a 25-mL sample size and the compounds divided into two groups to minimize coelution.

Method 524.2 – *Table 3*

Ion Abundance Criteria for 4-Bromofluorobenzene (BFB)

Mass (M/z)	Relative Abundance Criteria
50	15-40% of mass 95
75	30-80% of mass 95
95	Base Peak, 100% Relative Abundance
96	5-9% of mass 95
173	<2% of mass 174
174	>50% of mass 95
175	5-9% of mass 174
176	> 95% but < 101% of mass 174
177	5-9% of mass 176

Method 524.2 – *Table 4*

Accuracy and Precision Data From 16-31
Determinations of the Method Analytes in Reagent Water Using Wide Bore
Capillary Column 1[a]

Compound	True Conc. Range, (µg/L)	Mean Accuracy (% of true value)	Rel. Std. Dev. (%)	Method Det. Limit (µg/L)
Benzene	0.1 - 10	97	5.7	0.04
Bromobenzene	0.1 - 10	100	5.5	0.03
Bromochloromethane	0.5 - 10	90	6.4	0.04
Bromodichloromethane	0.1 - 10	95	6.1	0.08
Bromoform	0.5 - 10	101	6.3	0.12
Bromomethane	0.5 - 10	95	8.2	0.11
n-Butylbenzene	0.5 - 10	100	7.6	0.11
sec-Butylbenzene	0.5 - 10	100	7.6	0.13
tert-Butylbenzene	0.5 - 10	102	7.3	0.14
Carbon tetrachloride	0.5 - 10	84	8.8	0.21
Chlorobenzene	0.1 - 10	98	5.9	0.04
Chloroethane	0.5 - 10	89	9.0	0.10
Chloroform	0.5 - 10	90	6.1	0.03
Chloromethane	0.5 - 10	93	8.9	0.13
2-Chlorotoluene	0.1 - 10	90	6.2	0.04
4-Chlorotoluene	0.1 - 10	99	8.3	0.06
Dibromochloromethane	0.1 - 10	92	7.0	0.05
1,2-Dibromo-3-chloropropane	0.5 - 10	83	19.9	0.26
1,2-Dibromoethane	0.5 - 10	102	3.9	0.06
Dibromomethane	0.5 - 10	100	5.6	0.24
1,2-Dichlorobenzene	0.1 - 10	93	6.2	0.03
1,3-Dichlorobenzene	0.5 - 10	99	6.9	0.12
1,4-Dichlorobenzene	0.2 - 20	103	6.4	0.03
Dichlorodifluoromethane	0.5 - 10	90	7.7	0.10
1,1-Dichloroethane	0.5 - 10	96	5.3	0.04
1,2-Dichloroethane	0.1 - 10	95	5.4	0.06
1,1-Dichloroethene	0.1 - 10	94	6.7	0.12
cis-1,2-Dichloroethene	0.5 - 10	101	6.7	0.12
trans-1,2-Dichloroethene	0.1 - 10	93	5.6	0.06
1,2-Dichloropropane	0.1 - 10	97	6.1	0.04
1,3-Dichloropropane	0.1 - 10	96	6.0	0.04
2,2-Dichloropropane	0.5 - 10	86	16.9	0.35
1,1-Dichloropropene	0.5 - 10	98	8.9	0.10
cis-1,2-Dichloropropene				
trans-1,2-Dichloropropene				
Ethyl benzene	0.1 - 10	99	8.6	0.06
Hexachlorobutadiene	0.5 - 10	100	6.8	0.11
Isopropylbenzene	0.5 - 10	101	7.6	0.15

Method 524.2 – *Table 4*

Accuracy and Precision Data From 16-31
Determinations of the Method Analytes in Reagent Water Using Wide Bore
Capillary Column 1[a] (Continued)

Compound	True Conc. Range, (ug/L)	Mean Accuracy (% of true value)	Rel. Std. Dev. (%)	Method Det. Limit (ug/L)
p-Isopropyltoluene (4-Isopropyl toluene)	0.1 - 10	99	6.7	0.12
Methylene chloride	0.1 - 10	95	5.3	0.03
Naphthalene	0.1 - 100	104	8.2	0.04
n-Propylbenzene	0.1 - 10	100	5.8	0.04
Styrene	0.1 - 100	102	7.2	0.04
1,1,1,2-Tetrachloroethane	0.5 - 10	90	6.8	0.05
1,1,2,2-Tetrachloroethane	0.1 - 10	91	6.3	0.04
Tetrachloroethene	0.5 - 10	89	6.8	0.14
Toluene	0.5 - 10	102	8.0	0.11
1,2,3-Trichlorobenzene	0.5 - 10	109	8.6	0.03
1,2,4-Trichlorobenzene	0.5 - 10	108	8.3	0.04
1,1,1-Trichloroethane	0.5 - 10	98	8.1	0.08
1,1,2-Trichloroethane	0.5 - 10	104	7.3	0.10
Trichloroethene	0.5 - 10	90	7.3	0.19
Trichlorofluoromethane	0.5 - 10	89	8.1	0.08
1,2,3-Trichloropropane	0.5 - 10	108	14.4	0.32
1,2,4-Trimethylbenzene	0.5 - 10	99	8.1	0.13
1,3,5-Trimethylbenzene	0.5 - 10	92	7.4	0.05
Vinyl chloride	0.5 - 10	98	6.7	0.17
o-Xylene	0.1 - 31	103	7.2	0.11
m-Xylene	0.1 - 10	97	6.5	0.05
p-Xylene	0.5 - 10	104	7.7	0.13

[a] Data obtained by Robert W. Slater using column 1 (60m x 0.75mm ID with 1.5 um film thickness VOCOL (Supelco, Inc.)), with a jet separator interface and a quadrupole mass spectrometer (sect. 11.3.1) with analytes divided among three solutions.

Method 524.2 – *Table 5*

Accuracy and Precision Data from Seven Determinations of the Method Analytes in Reagent Water Using a Narrow Bore Capillary Column 3[a]

Compound	True Conc. Range, (µg/L)	Mean Accuracy (% of true value)	Rel. Std. Dev. (%)	Method Det. Limit (µg/L)
Benzene	0.1	99	6.2	0.03
Bromobenzene	0.5	97	7.4	0.11
Bromochloromethane	0.5	97	5.8	0.07
Bromodichloromethane	0.1	100	4.6	0.03
Bromoform	0.1	99	5.4	0.20
Bromomethane	0.1	99	7.1	0.06
n-Butylbenzene	0.5	94	6.0	0.03
sec-Butylbenzene	0.5	90	7.1	0.12
tert-Butylbenzene	0.5	90	2.5	0.33
Carbon tetrachloride	0.1	92	6.8	0.08
Chlorobenzene	0.1	91	5.8	0.03
Chloroethane	0.1	100	5.8	0.02
Chloroform	0.1	95	3.2	0.02
Chloromethane	0.1	99	4.7	0.05
2-Chlorotoluene	0.1	99	4.6	0.05
4-Chlorotoluene	0.1	96	7.0	0.05
Cyanogen chloride[b]		92	10.6	0.30
Dibromochloromethane	0.1	99	5.6	0.07
1,2-Dibromo-3-chloropropane	0.1	92	10.0	0.05
1,2-Dibromoethane	0.1	97	5.6	0.02
Dibromomethane	0.1	93	6.9	0.03
1,2-Dichlorobenzene	0.1	97	3.5	0.05
1,3-Dichlorobenzene	0.1	99	6.0	0.05
1,4-Dichlorobenzene	0.1	93	5.7	0.04
Dichlorodifluoromethane	0.1	99	8.8	0.11
1,1-Dichloroethane	0.1	98	6.2	0.03
1,2-Dichloroethane	0.1	100	6.3	0.02
1,1-Dichloroethene	0.1	95	9.0	0.05
cis-1,2-Dichloroethene	0.1	100	3.7	0.06
trans-1,2-Dichloroethene	0.1	98	7.2	0.03
1,2-Dichloropropane	0.1	96	6.0	0.02
1,3-Dichloropropane	0.1	99	5.8	0.04
2,2-Dichloropropane	0.1	99	4.9	0.05
1,1-Dichloropropene	0.1	98	7.4	0.02
cis-1,2-Dichloropropene				
trans-1,2-Dichloropropene				
Ethyl benzene	0.1	99	5.2	0.03
Hexachlorobutadiene	0.1	100	6.7	0.04
Isopropylbenzene	0.5	98	6.4	0.10

Method 524.2 – *Table 5*

Accuracy and Precision Data from Seven Determinations of the Method Analytes In Reagent Water Using a Narrow Bore Capillary Column 3[a]

Compound	True Conc. Range, (μg/L)	Mean Accuracy (% of true value)	Rel. Std. Dev. (%)	Method Det. Limit (μg/L)
p-Isopropyltoluene (4-Isopropyl toluene)	0.5	87	13.0	0.26
Methylene chloride	0.5	97	13.0	0.09
Naphthalene	0.1	98	7.2	0.04
n-Propylbenzene	0.1	99	6.6	0.06
Styrene	0.1	96	19.0	0.06
1,1,1,2-Tetrachloroethane	0.1	100	4.7	0.04
1,1,2,2-Tetrachloroethane	0.1	100	12.0	0.20
Tetrachloroethene	0.1	96	5.0	0.05
Toluene	0.1	100	5.9	0.08
1,2,3-Trichlorobenzene	0.1	98	8.9	0.04
1,2,4-Trichlorobenzene	0.1	91	16.0	0.20
1,1,1-Trichloroethane	0.1	100	4.0	0.04
1,1,2-Trichloroethane	0.5	98	4.9	0.03
Trichloroethene	0.1	96	2.0	0.02
Trichlorofluoromethane	0.1	97	4.6	0.07
1,2,3-Trichloropropane	0.1	96	6.5	0.03
1,2,4-Trimethylbenzene	0.1	96	6.5	0.04
1,3,5-Trimethylbenzene	0.1	99	4.2	0.02
Vinyl chloride	0.1	96	0.2	0.04
o-Xylene	0.1	94	7.5	0.06
m-Xylene	0.1	94	4.6	0.03
p-Xylene	0.1	97	6.1	0.06

[a] Data obtained by Caroline A. Madding using column 3 (30m x 0.53mm ID DB-5 (J&W Scientific, Inc.)with a 1μm film thickness), with a cryogenic interface and a quadrupole mass spectrometer (sect. 11.3.3).

[b] Flesch, J.J., P.S. Fair, "The Analysis of Cyanogen Chloride in Drinking Water," Proceeding of Water Quality Technology Conference, American Water Works Association, St. Louis, MO., November 14-16, 1988.

Method 525.1 – *Table 1*

Ion Abundance Criteria for Bis(Perfluorophenyl)Phenyl Phosphine (Decafluorotriphenylphosphine, DFTPP)

Mass (M/z)	Relative Abundance Criteria	Purpose of Checkpoint[1]
51	10-80% of the base peak	low mass sensitivity
68	<2% of mass 69	low mass resolution
70	<2% of mass 69	low mass resolution
127	10-80% of the base peak	low-mid mass sensitivity
197	<2% of mass 198	mid-mass resolution
198	base peak or >50% of 442	mid-mass resolution and sensitivity
199	5-9% of mass 198	mid-mass resolution and isotope ratio
275	10-60% of the base peak	mid-high mass sensitivity
365	>1% of the base peak	baseline threshold
441	Present and < mass 443	high mass resolution
442	base peak or >50% of 198	high mass resolution and sensitivity
443	15-24% of mass 442	high mass resolution and isotope ratio

[1] All ions are used primarily to check the mass measuring accuracy of the mass spectrometer and data system, and this is the most important part of the performance test. The three resolution checks, which include natural abundance isotope ratios, constitute the next most important part of the performance test. The correct setting of the base line threshold, as indicated by the presence of low intensity ions, is the next most important part of the performance test. Finally, the ion abundance ranges are designed to encourage some standarization to fragmentation patterns.

Method 525.1 – *Table 2*

Retention Time Data, Quantitiation Ions, and Internal Standard References for Method Analytes

INTERNAL STANDARDS

Compound	Compound Number	Retention Time (min:sec)		Quantitation Ion (m/z)	Internal Standard Reference
		A[a]	B[b]		
Acenaphthene-d_{10}	1	4:49	7:45	164	--
Phenanthrene-d_{10}	2	8:26	11:08	188	--
Chrysene-d_{12}	3	18:14	19:20	240	--

SURROGATE

Compound	Compound Number	Retention Time (min:sec)		Quantitation Ion (m/z)	Internal Standard Reference
		A[a]	B[b]		
Perylene-d_{12}	4	23:37	22:55	264	3

INTERNAL STANDARDS

Compound	Compound Number	Retention Time (min:sec)		Quantitation Ion (m/z)	Internal Standard Reference
		A[a]	B[b]		
Acenaphthylene	5	4:37	7:25	152	1
Aldrin	6	11:21	13:36	66	2
Anthracene	7	8:44	11:20	178	2
Atrazine	8	7:56	10:42	200/215	1/2
Benzo(a)anthracene	9	18:06	19:14	228	3
Benzo(b)fluoranthene	10	22:23	22:07	252	3
Benzo(k)fluoranthene	11	22:28	22:07	252	3
Benzo(a)pyrene	12	22:28	22:47	252	3
Benzo(g,h,i)perylene	13	27:56	26:44	276	3
Butylbenzylphthalate	14	16:40	18:09	149	2/3
alpha-chlordane	15	13:44	15:42	375	2/3
gamma-chlordane	16	13:16	15:18	375	2/3
trans nonachlor	17	13:54	15:50	409	2/3
2-Chlorobiphenyl	18	4:56	7:55	188	1
Chrysene	19	18:24	19:23	228	3
Dibenz(a,h)anthracene	20	27:15	25:57	278	3
di-n-butylphthalate	21	10:58	13:20	149	2
2,3-Dichlorobiphenyl	22	7:20	10:12	222	1
Diethylphthalate	23	5:52	8:50	149	1
di(2-ethylhexyl)phthalate	24	19:19	20:01	149	2/3
di(2-ethylhexyl)adipate	25	17:17	18:33	129	2/3

Method 525.1 – *Table 2*

Retention Time Data, Quantitation Ions, and Internal Standard References for Method Analytes (Continued)

INTERNAL STANDARDS					
Compound	Compound Number	Retention Time (min:sec) A[a]	B[b]	Quantitation Ion (m/z)	Internal Standard Reference
Dimethylphthalate	26	4:26	7:21	163	1
Endrin	27	15:52	16:53	81	2/3
Fluorene	28	6:00	8:53	166	1
Heptachlor	29	10:20	12:45	100/160	2
Heptachlor epoxide	30	12:33	14:40	81/353	2
2,2',3,3',4,4',6-Heptachlorobiphenyl	31	18:25	19:25	394/396	3
Hexachlorobenzene	32	7:37	10:20	284/286	1/2
2,2',4,4',5,6'-Hexachlorobiphenyl	33	14:34	16:30	360	2
Hexachlorocyclopentadiene	34	3:36	6:15	237	1
Indeno(1,2,3,c,d)pyrene	35	27:09	25:50	276	3
Lindane	36	8:17	10:57	181/183	1/2
Methoxychlor	37	18:34	19:30	227	3
2,2',3,3',4,5',6,6'-Octachlorobiphenyl	38	18:38	19:33	430	3
2,2',3',4,6,-Pentachlorobiphenyl	39	12:50	15:00	326	2
Pentachlorophenol	40	8:11	10:51	266	2
Phenanthrene	41	8:35	11:13	178	2
Pyrene	42	13:30	15:29	202	2/3
Simazine	43	7:47	10:35	201	1/2
2,2',4,4'-Tetrachlorobiphenyl	44	11:01	13:25	292	2
Toxaphene	45	11:30-23:30	13:00-21:30	159	2
2,4,5-Trichlorobiphenyl	46	9:23	11:59	256	2
Alachlor	47	--	13:19	160	2

[a]   Single ramp linear temperature program conditions (sect. 9.2.3.2). Adjust the helium carrier gas flow rate to about 33 cm/sec. Inject at 40 °C and hold in splitless mode for 1 min. Heat rapidly to 160 °C. At 3 min start the temperature program: 160-320 °C at 6 °C/min; hold at 320 °C for 2 min. Start data acquisition at 3 min.

[b]   Multi-ramp linear temperature program conditions (sect. 9.2.3.1). Multi-ramp temperature program GC conditions. Adjust the helium carrier gas flow rate to about 33 cm/sec. Inject at 45 °C and hold in splitless mode for 1 min. Heat rapidly to 130 °C. At 3 min start the temperature program 130-180 °C at 12 °C/min; 180-240 °C at 7 °C/min; 240-320 °C at 12 °C/min. Start data acquisition at 5 min.

Method 525.1 – *Table 3*

Accuracy and Precision Data for Seven Determination of the Method Analytes at 0.2 µg/L with Liquid — Solid Extraction and the Ion Trap Mass Spectrometer

Compound Number[a]	True Conc. µg/L	Mean Observed Conc. (µg/L)	Std. Dev. (µg/L)	Rel. Std. Dev. (%)	Mean Method Accuracy (% of True Conc.)	Method Detection Limit (MDL) (µg/L)
4	5	5	0.3	6	100	a
5	2	1.9	0.2	11	95	a
6	2	1.6	0.2	13	80	a
7	2	1.7	0.1	5.9	85	a
8	2	2.2	0.3	14	110	a
9	2	1.8	0.2	11	90	a
10	2	not separated from No. 11; measured with No. 11				
11	2	4.2	0.3	7.1	105	a
12	2	0.8	0.2	25	40	a
13	2	0.7	0.1	14	35	a
14	2	2	0.3	15	100	a
15	2	2	0.2	10	100	a
16	2	2.2	0.3	14	110	a
17	2	2.7	1	37	135	a
18	2	1.9	0.1	5.2	95	a
19	2	2.2	0.1	4.5	110	a
20	2	0.3	0.3	100	15	a
21	2	2.2	0.3	14	110	a
22	2	2.3	0.1	4.3	115	a
23	2	2	0.3	15	100	a
24	2	1.9	0.2	11	95	a
25	2	1.6	0.3	19	80	a
26	2	1.9	0.2	11	95	a
27	2	1.8	0.1	5.5	90	a
28	2	2.2	0.2	9.1	110	a
29	2	2.2	0.3	14	110	a
30	2	2.3	0.2	8.7	115	a
31	2	1.4	0.2	14	70	a
32	2	1.7	0.2	12	85	a
33	2	1.6	0.4	25	80	a
34	2	1.1	0.1	9.1	55	a

Method 525.1 – *Table 3*

Accuracy and Precision Data for Seven Determination of the Method Analytes at 0.2 µg/L with Liquid — Solid Extraction and the Ion Trap Mass Spectrometer (Continued)

Compound Number[a]	True Conc. µg/L	Mean Observed Conc. (µg/L)	Std. Dev. (µg/L)	Rel. Std. Dev. (%)	Mean Method Accuracy (% of True Conc.)	Method Detection Limit (MDL) (µg/L)
35	2	0.4	0.2	50	20	a
36	2	2.1	0.2	9.5	105	a
37	2	1.8	0.2	11	90	a
38	2	1.8	0.2	11	90	a
39	2	1.9	0.1	5.3	95	a
40	8	8.2	1.2	15	102	a
41	2	2.4	0.1	4.2	120	a
42	2	1.9	0.1	5.3	95	a
43	2	2.1	0.2	9.5	105	a
44	2	1.5	0.1	6.7	75	a
45	25	28	4.7	17	112	15
46	2	1.7	0.1	5.9	85	a
Mean[b]	2	1.8	0.2	15	91	0.6

(a) See Table 2, page A-35 for compound names.

(b) Compounds 4, 40, 45 exclude from the means; See Table 4, page A-39

Method 525.1 – *Table 4*

Accuracy and Precision Data from Five to Seven Determinations of the Method Analytes at 0.2 μg/L with Liquid-Solid Extraction and the Ion Trap Mass Spectrometer

Compound Number[a]	True Conc. μg/L	Mean Observed Conc. (μg/L)	Std. Dev. (μg/L)	Rel. Std. Dev. (%)	Mean Method Accuracy (% of True Conc.)	Method Detection Limit (MDL) (μg/L)
4	0.5	0.45	0.6	13	90	0.1
5	0.2	0.13	0.03	23	65	0.1
6	0.2	0.13	0.03	23	65	0.1
7	0.2	0.13	0.01	7.7	65	0.04
8	0.2	0.24	0.03	13	120	0.1
9	0.2	0.14	0.01	7.1	70	0.04
10	0.2	not separated from No.11; measured with No. 11				
11	0.2	0.25	0.04	16	62	0.2
12	0.2	0.03	0.01	33	15	0.04
13	0.2	0.03	0.02	67	15	0.1
14	0.2	0.32	0.07	22	160	0.3
15	0.2	0.17	0.04	24	85	0.2
16	0.2	0.19	0.03	16	95	0.1
17	0.2	0.17	0.08	47	85	0.3
18	0.2	0.19	0.03	16	95	0.1
19	0.2	0.21	0.01	4.8	105	0.04
20	0.2	0.03	0.02	67	150	0.1
21	0.2	0.48	0.09	19	240	0.3
22	0.2	0.20	0.03	15	100	0.1
23	0.2	0.45	0.21	47	225	0.8
24	0.2	0.39	0.16	41	195	0.6
25	0.2	0.31	0.16	52	155	0.6
26	0.2	0.21	0.01	4.8	105	0.04
27	0.2	0.12	0.12	100	60	0.5
28	0.2	0.21	0.05	24	105	0.2
29	0.2	0.22	0.01	4.5	110	0.04
30	0.2	0.19	0.04	21	95	0.2
31	0.2	0.19	0.03	16	95	0.1
32	0.2	0.16	0.04	25	80	0.1
33	0.2	0.19	0.03	16	95	0.1
34	0.2	0.04	0.01	25	20	0.03
35	0.2	0.04	0.03	75	20	0.1
36	0.2	0.22	0.02	9.1	110	0.1
37	0.2	0.11	0.01	9.1	55	0.04
38	0.2	0.19	0.05	26	95	0.2

Method 525.1 – *Table 4*

Accuracy and Precision Data from Five to Seven Determinations of the Method Analytes at 0.2 µg/L with Liquid-Solid Extraction and the Ion Trap Mass Spectrometer (Continued)

Compound Number[a]	True Conc. µg/L	Mean Observed Conc. (µg/L)	Std. Dev. (µg/L)	Rel. Std. Dev. (%)	Mean Method Accuracy (% of True Conc.)	Method Detection Limit (MDL) (µg/L)
39	0.2	0.13	0.02	15	65	0.1
40	0.8	0.78	0.08	10	97	0.3
41	0.2	0.20	0.004	2.0	100	0.01
42	0.2	0.18	0.005	2.8	90	0.02
43	0.2	0.25	0.04	16	125	0.2
44	0.2	0.14	0.04	29	70	0.1
45		not measured at this level				
46	0.2	0.13	0.02	15	65	0.06
Mean[b]	0.2	0.18	0.04	25	95	0.16

(a) See Table 2, page A-35 for compound names.

(b) Compounds 4, 40, and 45 excluded from the means.

Method 525.1 – *Table 5*

Accuracy and Precision Data from Seven Determinations of the Method Analytes at 2 µg/L with Liquid-Solid Extraction and a Magnetic Sector Mass Spectrometer

Compound Number[a]	True Conc. µg/L	Mean Observed Conc. (µg/L)	Std. Dev. (µg/L)	Rel. Std. Dev. (%)	Mean Method Accuracy (% of True Conc.)	Method Detection Limit (MDL) (µg/L)
4	5	5.7	0.34	6.0	114	(b)
5	2	1.9	0.22	12	95	(b)
6	2	1.6	0.18	11	80	(b)
7	2	2.2	0.67	30	110	(b)
8	2	2.4	0.46	19	120	(b)
9	2	2.2	0.87	40	110	(b)
10	2	not separated from No. 11; measured with No. 11				
11	2	4.0	0.37	9.3	100	(b)
12	2	0.85	0.15	18	43	(b)
13	2	0.69	0.12	17	35	(b)
14	2	2.0	0.20	10	100	(b)
15	2	2.2	0.41	19	110	(b)
16	2	2.1	0.38	18	105	(b)
17	2	1.9	0.10	5.2	95	(b)
18	2	2.0	0.29	14	100	(b)
19	2	2.1	0.32	15	105	(b)
20	2	0.75	0.18	24	38	(b)
21	2	2.5	0.32	13	125	(b)
22	2	2.0	0.23	12	100	(b)
23	2	3.5	1.8	51	175	(b)
24	2	2.0	0.28	14	100	(b)
25	2	1.4	0.16	11	70	(b)
26	2	2.9	0.70	24	145	(b)
27	2	1.7	0.45	26	85	(b)
28	2	2.6	1.0	38	130	(b)
29	2	1.2	0.10	8.3	60	(b)
30	2	2.6	0.42	16	130	(b)
31	2	1.5	0.19	13	75	(b)
32	2	1.5	0.35	23	75	(b)
33	2	1.9	0.17	8.9	95	(b)
34	2	0.89	0.11	12	45	(b)
35	2	0.83	0.072	8.7	42	(b)
36	2	2.2	0.10	4.5	110	(b)
37	2	2.0	0.88	44	100	(b)
38	2	1.5	0.11	7.3	75	(b)

Method 525.1 – *Table 5*

Accuracy and Precision Data from Seven Determinations of the Method Analytes at 2 µg/L with Liquid-Solid Extraction and a Magnetic Sector Mass Spectrometer (Continued)

Compound Number[a]	True Conc. µg/L	Mean Observed Conc. (µg/L)	Std. Dev. (µg/L)	Rel. Std. Dev. (%)	Mean Method Accuracy (% of True Conc.)	Method Detection Limit (MDL) (µg/L)
39	2	1.6	0.14	8.8	80	(b)
40	8	12	2.6	22	150	(b)
41	2	2.3	0.18	7.8	115	(b)
42	2	2.0	0.26	13	100	(b)
43	2	2.5	0.34	14	125	(b)
44	2	1.6	0.17	11	80	(b)
45	25	28	2.7	10	112	9
46	2	1.9	0.073	3.8	95	(b)
Mean[c]	2	1.8	0.32	16.0	88	1

(a) See Table 2, page A-35 for compound names.

(b) See Table 6.

(c) Compounds 4, 40, and 45 excluded from the means.

Method 525.1 – *Table 6*

Accuracy and Precision Data from Six or Seven Determinations of the Method Analytes at 0.2 µg/L with Liquid-Solid Extraction and a Magnetic Sector Mass Spectrometer

Compound Number(a)	True Conc. µg/L	Mean Observed Conc. (µg/L)	Std. Dev. (µg/L)	Rel. Std. Dev. (%)	Mean Method Accuracy (% of True Conc.)	Method Detection Limit (MDL) (µg/L)
4	0.5	0.67	0.07	9.4	134	0.2
5	0.2	0.11	0.03	24	55	0.1
6	0.2	0.11	0.02	21	56	0.1
7	0.2	0.14	0.02	17	70	0.1
8	0.2	0.26	0.08	31	130	0.3
9	0.2	0.24	0.06	26	120	0.2
10	0.2	not separated from No.11; measured with No. 11				
11	0.2	0.40	0.10	25	100	0.3
12	0.2	0.08	0.02	27	38	0.1
13	0.2	0.07	0.01	22	33	0.1
14	0.2	0.33	0.16	48	160	0.5
15	0.2	0.19	0.02	13	95	0.1
16	0.2	0.17	0.08	45	85	0.3
17	0.2	0.19	0.04	18	95	0.1
18	0.2	0.17	0.02	13	85	0.1
19	0.2	0.27	0.08	28	135	0.3
20	0.2	0.09	0.01	15	46	0.1
21	0.2	1.1	1.2	109	550	4
22	0.2	0.18	0.05	30	90	0.2
23	0.2	0.29	0.17	59	145	0.6
24	0.2	0.42	0.23	55	210	0.8
25	0.2	0.32	0.16	50	160	0.5
26	0.2	0.20	0.09	47	100	0.3
27	0.2	0.53	0.30	57	265	1
28	0.2	0.18	0.03	15	90	0.1
29	0.2	0.11	0.05	42	55	0.2
30	0.2	0.33	0.08	26	165	0.3
31	0.2	0.17	0.01	7.1	85	0.04
32	0.2	0.11	0.04	40	55	0.2
33	0.2	0.17	0.03	15	85	0.1
34	0.2	0.05	0.02	35	24	0.1
35	0.2	0.08	0.06	8.1	40	0.02
36	0.2	0.27	0.03	11	135	0.1

Method 525.1 – *Table 6*

Accuracy and Precision Data from Six or Seven Determinations of the Method Analytes at 0.2 μg/L with Liquid-Solid Extraction and a Magnetic Sector Mass Spectrometer (Continued)

Compound Number[a]	True Conc. μg/L	Mean Observed Conc. (μg/L)	Std. Dev. (μg/L)	Rel. Std. Dev. (%)	Mean Method Accuracy (% of True Conc.)	Method Detection Limit (MDL) (μg/L)
37	0.2	0.24	0.09	39	120	0.3
38	0.2	0.15	0.02	12	75	0.1
39	0.2	0.13	0.02	13	65	0.1
40	0.8	1.8	0.82	46	225	3
41	0.2	0.21	0.07	33	105	0.2
42	0.2	0.19	0.04	23	95	0.1
43	0.2	0.27	0.07	27	135	0.2
44	0.2	0.13	0.03	22	65	0.1
45		not measured at this level				
46	0.2	0.16	0.04	23	80	0.12
Mean[b]	0.2	0.21	0.09	28	102	0.3

(a) See Table 2, page A-35 for compound names.

(b) Compounds 4, 40, 45 excluded from the means.

Method 525.1 – *Table 7*

Accuracy and Precision Data from Seven Determinations at 2.0 µg/L with Liquid-Solid Extraction and a Quadrupole Mass Spectrometer

Compound Number(a)	True Conc. µg/L	Mean Observed Conc. (µg/L)	Std. Dev. (µg/L)	Rel. Std. Dev. (%)	Mean Method Accuracy (% of True Conc.)	Method Detection Limit (MDL) (µg/L)
47	2	2.4	0.4	16.	122	1.0

(a) See Table 2, page A-35 for compound names.

Method 525.1 – *Table 10*

Minimum Detection Limits from Seven Replicates using Liquid-Solid Extraction C-18 Disks and an Ion Trap Mass Spectrometer

Chemical Name	CAS Number	Minimum Detection Limits
Acenaphthylene	208-96-8	0.033
Alachlor	15972-60-8	0.092
Aldrin	309-00-2	0.083
Anthracene	120-12-7	0.086
Atrazine	1912-24-9	0.140
Benzo(a)anthracene	56-55-3	0.224
Benzo(b)fluoranthene	205-99-2	0.488
Benzo(k)fluoranthene	207-08-9	0.086
Benzo(a)pyrene	50-32-8	0.137
Benzo(g,h,i)perylene	191-24-2	0.094
Butylbenzylphthalate[1]	85-68-7	0.204
Chlordane-alpha	5103-71-9	0.384
Chlordane-gamma	5103-74-2	0.200
Chlordane (*trans*-Nonachlor-)	39765-80-5	0.574
Chrysene	218-01-9	0.068
Dibenz(a,h)anthracene	53-70-3	0.144
di-*n*-Butylphthalate	84-74-2	0.253
Diethylphthalate	84-66-2	0.075
di(2-Ethylhexyl)phthalate (bis(2-Ethylhexyl)phthalate)	117-81-7	1.584
di(2-Ethylhexyl)adipate	103-23-1	0.131
Dimethylphthalate	131-11-3	0.048
Endrin	72-20-8	0.160
Fluorene	86-73-7	0.046
Heptachlor	76-44-8	0.144
Heptachlor epoxide	1024-57-3	0.244
Hexachlorobenzene	118-74-1	0.111
Hexachlorocyclopentadiene	77-47-4	0.039
Indeno(1,2,3-cd)pyrene	193-39-5	0.170
Lindane (HCH-g)	58-89-9	0.041
Methoxychlor	72-43-5	0.048
PCB-mono-Cl-isomer (2-Chlorobiphenyl)	2051-60-7	0.045
PCB-di-Cl-isomer (2,3-Dichlorobiphenyl)	16605-91-7	0.061
PCB-tri-Cl-isomer (2,4,5-Trichlorobiphenyl)	15862-07-4	0.135
PCB-tetra-Cl-isomer (2,2',4,4'-Tetrachlorobiphenyl)	2437-79-8	0.177
PCB-penta-Cl-isomer (2,2',3',4,6-Pentachlorobiphenyl)	60233-25-2	0.095
PCB-hexa-Cl-isomer (2,2',4,4',5,6'-Hexachlorobiphenyl)	60145-22-4	0.200
PCB-hepta-Cl-isomer (2,2',3,3',4,4',6-Heptachlorobiphenyl)	52663-71-5	0.239
PCB-octa-Cl-isomer (2,2',3,3',4,5',6,6'-Octachlorobiphenyl)	40186-71-8	0.133

Method 525.1 – *Table 10*

Minimum Detection Limits from Seven Replicates using Liquid-Solid Extraction C-18 Disks and an Ion Trap Mass Spectrometer (Continued)

Chemical Name	CAS Number	Minimum Detection Limits
Pentachlorophenol	87-86-5	47.648
Phenanthrene	85-01-8	0.076
Pyrene	129-00-0	0.064
Simazine	122-34-9	0.118
Toxaphene	8001-35-2	7.763

[1]Some methods list as Benzyl butyl phthalate

Method 601 – *Table 1*

Chromatographic Conditions and Method Detection Limits

| Compound | Retention Time (min) | | Method Detection |
	Col.1	Col.2	Limit (µg/L)
Chloromethane	1.50	5.28	0.08
Bromomethane	2.17	7.05	1.18
Dichlorodifluoromethane	2.62	nd	1.81
Vinyl chloride	2.67	5.28	0.18
Chloroethane	3.33	8.68	0.52
Methylene chloride	5.25	10.1	0.25
Trichlorofluoromethane	7.18	nd	nd
1,1-Dichloroethene	7.93	7.72	0.13
1,1-Dichloroethane	9.30	12.6	0.07
trans-1,2-Dichloroethene	10.1	9.38	0.10
Chloroform	10.7	12.1	0.05
1,2-Dichloroethane	11.4	15.4	0.03
1,1,1-Trichloroethane	12.6	13.1	0.03
Carbon tetrachloride	13.0	14.4	0.12
Bromodichloromethane	13.7	14.6	0.10
1,2-Dichloropropane	14.9	16.6	0.04
cis-1,3-Dichloropropene	15.2	16.6	0.34
Trichloroethene	15.8	13.1	0.12
Dibromochloromethane	16.5	16.6	0.09
1,1,2-Trichloroethane	16.5	18.1	0.02
trans-1,3-Dichloropropene	16.5	18.0	0.20
2-Chloroethylvinyl ether	18.0	nd	0.13
Bromoform	19.2	19.2	0.20
1,1,2,2-Tetrachloroethane	21.6	nd	0.03
Tetrachloroethene	21.7	15.0	0.03
Chlorobenzene	24.2	18.8	0.25
1,3-Dichlorobenzene	34.0	22.4	0.32
1,2-Dichlorobenzene	34.9	23.5	0.15
1,4-Dichlorobenzene	35.4	22.3	0.24

Column 1 conditions: Carbopack B (60/80 mesh) coated with 1% SP-1000 packed in an 8 ft x 0.1 in. ID stainless steel or glass column with helium carrier gas at 40 mL/min flow rate. Column temperature held at 45 °C for 3 min then programmed at 8 °C/min to 220 °C and held for 15 min.

Column 2 conditions: Porisil-C (100/120 mesh) coated with n-octane packed in a 6 ft x 0.1 in. ID stainless steel or glass column with helium carrier gas at 40 mL/min flow rate. Column temperature held at 50 °C for 3 min then programmed a 6 °C/min to 170 °C and held for 4 min.

nd = not determined

Method 601 - *Table 2*

Calibration and QC Acceptance Criteria - Method 601[a]

Parameter	Range for Q (µg/L)	Limit for s (µg/L)	Range for x̄ (µg/L)	Range for P, Ps (%)
Bromodichloromethane	15.2-24.8	4.3	10.7-32.0	42-172
Bromoform	14.7-25.3	4.7	5.0-29.3	13-159
Bromomethane	11.7-28.3	7.6	3.4-24.5	D-144
Carbon tetrachloride	13.7-26.3	5.6	11.8-25.3	43-143
Chlorobenzene	14.4-25.6	5.0	10.2-27.4	38-150
Chloroethane	15.4-24.6	4.4	11.3-25.2	46-137
2-Chloroethylvinyl ether	12.0-28.0	8.3	4.5-35.5	14-186
Chloroform	15.0-25.0	4.5	12.4-24.0	49-133
Chloromethane	11.9-28.1	7.4	D-34.9	D-193
Dibromochloromethane	13.1-26.9	6.3	7.9-35.1	24-191
1,2-Dichlorobenzene	14.0-26.0	5.5	1.7-38.9	D-208
1,3-Dichlorobenzene	9.9-30.1	9.1	6.2-32.6	7-187
1,4-Dichlorobenzene	13.9-26.1	5.5	11.5-25.5	42-143
1,1-Dichloroethane	16.8-23.2	3.2	11.2-24.6	47-132
1,2-Dichloroethane	14.3-25.7	5.2	13.0-26.5	51-147
1,1-Dichloroethene	12.6-27.4	6.6	10.2-27.3	28-167
trans-1,2-Dichloroethene	12.8-27.2	6.4	11.4-27.1	38-155
1,2-Dichloropropane	14.8-25.2	5.2	10.1-29.9	44-156
cis-1,3-Dichloropropene	12.8-27.2	7.3	6.2-33.8	22-178
trans-1,3-Dichloropropene	12.8-27.2	7.3	6.2-33.8	22-178
Methylene chloride	15.5-24.5	4.0	7.0-27.6	25-162
1,1,2,2-Tetrachloroethane	9.8-30.2	9.2	6.6-31.8	8-184
Tetrachloroethene	14.0-26.0	5.4	8.1-29.6	26-162
1,1,1-Trichloroethane	14.2-25.8	4.9	10.8-24.8	41-138
1,1,2-Trichloroethane	15.7-24.3	3.9	9.6-25.4	39-136
Trichloroethene	15.4-24.6	4.2	9.2-26.6	35-146
Trichlorofluoromethane	13.3-26.7	6.0	7.4-28.1	21-156
Vinyl chloride	13.7-26.3	5.7	8.2-29.9	28-163

Q = Concentration measured in QC check sample, in µg/L (Section 7.5.3)

s = Standard deviation of four recovery measurements, in µg/L (Section 8.2.4)

x̄ = Average recovery for four recovery measurements, in µg/L (Section 8.2.4)

P, Ps = Percent recovery measured (Section 8.3.2, Section 8.4.2)

D = Detected; result must be greater than zero.

[a] Criteria were calculated assuming a QC check sample concentration of 20 µg/L.

Note: These criteria are based directly upon the method performance data in Table 3. Where necessary, the limits for recovery have been broadened to assure applicability of the limits to concentrations below those used to develop Table 3.

Method 602 – *Table 1*

Chromatographic Conditions and Method Detection Limits

Compound	Retention time (min)		Method Detection Limit (µg/L)
	Col.1	Col.2	
Benzene	3.33	2.75	0.2
Toluene	5.75	4.25	0.2
Ethylbenzene	8.25	6.25	0.2
Chlorobenzene	9.17	8.02	0.2
1,4-Dichlorobenzene	16.8	16.2	0.3
1,3-Dichlorobenzene	18.2	15	0.4
1,2-Dichlorobenzene	25.9	19.4	0.4

Column 1 conditions: Supelcoport (100/120 mesh) coated with 5% SP-1200/1.75% Bentone-34 packed in a 6 ft X 0.085 in. ID stainless steel column with helium carrier gas at 36 mL/min flow rate. Column temperature held at 50 °C for 2 min then programmed at 6 °C/min to 90 °C for a final hold.

Column 2 conditions: Chromosorb W-AW (60/80 mesh) coated with 5% 1,2,3-Tris(2-cyanoethyoxy)propane packed in a 6 ft X 0.085 in. ID stainless steel column with helium carrier gas at 30 mL/min flow rate. Column temperature held at 40 °C for 2 min then programmed at 2 °C/min to 100 °C for a final hold.

Method 602 – *Table 2*

Calibration and QC Acceptance Criteria - Method 602[a]

Parameter	Range for Q (µg/L)	Limit for s (µg/L)	Range for x̄ (µg/L)	Range for P, P_s (%)
Benzene	15.4-24.6	4.1	10.0-27.9	39-150
Chlorobenzene	16.1-23.9	3.5	12.7-25.4	55-135
1,2-Dichlorobenzene	13.6-26.4	5.8	10.6-27.6	37-154
1,3-Dichlorobenzene	14.5-25.5	5.0	12.8-25.5	50-141
1,4-Dichlorobenzene	13.9-26.1	5.5	11.6-25.5	42-143
Ethylbenzene	12.6-27.4	6.7	10.0-28.2	32-160
Toluene	15.5-24.5	4.0	11.2-27.7	46-148

Q   =   Concentration measured in QC check sample, in µg/L (Section 7.5.3)

s   =   Standard deviation of four recovery measurements, in µg/L (Section 8.2.4)

x̄   =   Average recovery for four recovery measurements, in µg/L (Section 8.2.4)

P, P_s =   Percent recovery measured (Section 8.3.2, Section 8.4.2)

[a] Criteria were calculated assuming a QC check sample concentration of 20 µg/L.

Note: These criteria are based directly upon the method performance data in Table 3. Where necessary, the limits for recovery have been broadened to assure applicability of the limits to concentrations below those used to develop Table 3.

Method 608 – *Table 1*

Chromatographic Conditions and Method Detection Limits

Parameter	Retention Time (min) Col.1	Col.2	Method Detection Limit (μg/L)
α-BHC	1.35	1.82	0.003
γ-BHC (Lindane)	1.70	2.13	0.004
β-BHC	1.90	1.97	0.006
Heptachlor	2.00	3.35	0.003
δ-BHC	2.15	2.20	0.009
Aldrin	2.40	4.10	0.004
Heptachlor epoxide	3.50	5.00	0.083
Endosulfan I	4.50	6.20	0.014
4,4'-DDE	5.13	7.15	0.004
Dieldrin	5.45	7.23	0.002
Endrin	6.55	8.10	0.006
4,4'-DDD	7.83	9.08	0.011
Endosulfan II	8.00	8.28	0.004
4,4'-DDT	9.40	11.75	0.012
Endrin aldehyde	11.82	9.30	0.023
Endosulfan sulfate	14.22	10.70	0.066
Chlordane	mr	mr	0.014
Toxaphene	mr	mr	0.24
PCB-1016	mr	mr	nd
PCB-1221	mr	mr	nd
PCB-1232	mr	mr	nd
PCB-1242	mr	mr	0.065
PCB-1248	mr	mr	nd
PCB-1254	mr	mr	nd
PCB-1260	mr	mr	nd

Column 1 conditions: Supelcoport (100/120 mesh) coated with 1.5% SP-2250/1.95% SP-2401 packed in a 1.8 m long X 4 mm ID glass column with 5% methane/95% argon carrier gas at 60 mL/min flow rate. Column temperature held isothermal at 200 °C, except for PCB-1016 through PCB-1248, should be measured at 160 °C.

Column 2 conditions: Supelcoport (100/120 mesh) coated with 3% OV-1 packed in a 1.8 m long X 4 mm ID glass column with 5% methane/95% argon carrier gas at 60 mL/min flow rate. Column temperature held isothermal at 200 °C for the pesticides; at 140 °C for PCB-1221 and 1232; and at 170 °C for PCB-1016 and 1242 to 1268.

mr = Multiple peak response

nd = not determined

Method 608 -- *Table 3*

QC Acceptance Criteria

Parameter	Test conc. (μg/L)	Limit for s (μg/L)	Range for $\bar{x}$ (μg/L)	Range for P, P_s (%)
Aldrin	2.0	0.042	1.08-2.24	42-122
α-BHC	2.0	0.48	0.98-2.44	37-134
β-BHC	2.0	0.64	0.78-2.60	17-147
δ-BHC	2.0	0.72	1.01-2.37	19-140
γ-BHC	2.0	0.46	0.86-2.32	32-127
Chlordane	50	10.0	27.6-54.3	45-119
4,4'-DDD	10	2.8	4.8-12.6	31-141
4,4'-DDE	2.0	0.55	1.08-2.60	30-145
4,4'-DDT	10	3.6	4.6-13.7	25-160
Dieldrin	2.0	0.76	1.15-2.49	36-146
Endosulfan I	2.0	0.49	1.14-2.82	45-153
Endosulfan II	10	6.1	2.2-17.1	D-202
Endosulfan sulfate	10	2.7	3.8-13.2	26-144
Endrin	10	3.7	5.1-12.6	30-147
Heptachlor	2.0	0.4	0.86-2.00	34-111
Heptachlor epoxide	2.0	0.41	1.13-2.63	37-142
Toxaphene	50	12.7	27.8-55.6	41-126
PCB-1016	50	10.0	30.5-51.5	50-114
PCB-1221	50	24.4	22.1-75.2	15-178
PCB-1232	50	17.9	14.0-98.5	10-215
PCB-1242	50	12.2	24.8-69.6	39-150
PCB-1248	50	15.9	29.0-70.2	38-158
PCB-1254	50	13.8	22.2-57.9	29-131
PCB-1260	50	10.4	18.7-54.9	8-127

s   =   Standard deviation of four recovery measurements, in μg/L (Section 8.2.4)

$\bar{x}$ =   Average recovery for four recovery measurements, in μg/L (Section 8.2.4)

P, P_s =   Percent recovery measured (Section 8.3.2, Section 8.4.2)

D   =   Detected; result must be greater than zero.

Note: These criteria are based directly upon the method performance data in Table 4. Where necessary, the limits for recovery have been broadened to assure applicability of the limits to concentrations below those used to develop Table 4.

Method 624 – *Table 1*

Chromatographic Conditions and Method Detection Limits

Parameter	Retention Time (min)	Method Detection Limit (µg/L)
Chloromethane	2.3	nd
Bromomethane	3.1	nd
Vinyl chloride	3.8	nd
Chloroethane	4.6	nd
Methylene chloride	6.4	2.8
Trichlorofluoromethane	8.3	nd
1,1-Dichloroethene	9.0	2.8
1,1-Dichloroethane	10.1	4.7
trans-1,2-Dichloroethene	10.8	1.6
Chloroform	11.4	1.6
1,2-Dichloroethane	12.1	2.8
1,1,1-Trichloroethane	13.4	3.8
Carbon tetrachloride	13.7	2.8
Bromodichloromethane	14.3	2.2
1,2-Dichloropropane	15.7	6.0
cis-1,3-Dichloropropene	15.9	5.0
Trichloroethene	16.5	1.9
Benzene	17.0	4.4
Dibromochloromethane	17.1	3.1
1,1,2-Trichloroethane	17.2	5.0
trans-1,3-Dichloropropene	17.2	nd
2-Chloroethylvinyl ether	18.6	nd
Bromoform	19.8	4.7
1,1,2,2-Tetrachloroethane	22.1	6.9
Tetrachloroethene	22.2	4.1
Toluene	23.5	6.0
Chlorobenzene	24.6	6.0
Ethyl benzene	26.4	7.2
1,3-Dichlorobenzene	33.9	nd
1,2-Dichlorobenzene	35.0	nd
1,4-Dichlorobenzene	35.4	nd

Column conditions: Carbopak B (60/80 mesh) coated with 1% SP-1000 packed in a 6 ft by .1 in. ID glass column with helium carrier gas at 30 mL/min. flow rate. Column temperature held at 45 °C for 3 min., then programmed at 8 °C/ min. to 220 °C and held for 15 min.

nd = not determined

Method 624 – *Table 2*

BFB Key M/Z Abundance Criteria

Mass	m/z Abundance criteria
50	15 to 40% of mass 95
75	30 to 60% of mass 95
95	Base Peak, 100% Relative Abundance
96	5 to 9% of mass 95
173	less than 2% of mass 174
174	greater than 50% of mass 95
175	5 to 9% of mass 174
176	greater than 95% but less than 101% of mass 174
177	5 to 9% of mass 176

Method 624 – *Table 3*

Suggested Surrogate and Internal Standards

Compound	Retention Time (min)[a]	Primary m/z	Secondary masses
Benzene d_6	17.0	84	
4-Bromofluorobenzene	28.3	95	174, 176
1,2-Dichloroethane d_4	12.1	102	
1,4-Difluorobenzene	19.6	114	63, 88
Ethylbenzene d_5	26.4	111	
Ethylbenzene d_{10}	26.4	98	
Fluorobenzene	18.4	96	70
Pentafluorobenzene	23.5	168	
Bromochloromethane	9.3	128	49, 130, 51
2-Bromo-1-chloropropane	19.2	77	79, 156
1,4-Dichlorobutane	25.8	55	90, 92

[a] For chromatographic conditions, see Table 1.

Method 624 – *Table 5*

Calibration and QC Acceptance Criteria[a]

Parameter	Range for Q (μg/L)	Limit for s (μg/L)	Range for $\bar{x}$ (μg/L)	Range for P, P_s (%)
Benzene	12.8-27.2	6.9	15.2-26.0	37-151
Bromodichloromethane	13.1-26.9	6.4	10.1-28.0	35-155
Bromoform	14.2-25.8	5.4	11.4-31.1	45-169
Bromomethane	2.8-37.2	17.9	D-41.2	D-242
Carbon tetrachloride	14.6-25.4	5.2	17.2-23.5	70-140
Chlorobenzene	13.2-26.8	6.3	16.4-27.4	37-160
Chloroethane	7.6-32.4	11.4	8.4-40.4	14-230
2-Chloroethylvinyl ether	D-44.8	25.9	D-50.4	D-305
Chloroform	13.5-26.5	6.1	13.7-24.2	51-138
Chloromethane	D-40.8	19.8	D-45.9	D-273
Dibromochloromethane	13.5-26.5	6.1	13.8-26.6	53-149
1,2-Dichlorobenzene	12.6-27.4	7.1	11.8-34.7	18-190
1,3-Dichlorobenzene	14.6-25.4	5.5	17.0-28.8	59-156
1,4-Dichlorobenzene	12.6-27.4	7.1	11.8-34.7	18-190
1,1-Dichloroethane	14.5-25.5	5.1	14.2-28.5	59-155
1,2-Dichloroethane	13.6-26.4	6.0	14.3-27.4	49-155
1,1-Dichloroethene	10.1-29.9	9.1	3.7-42.3	D-234
trans-1,2-Dichloroethene	13.9-26.1	5.7	13.6-28.4	54-156
1,2-Dichloropropane	6.8-33.2	13.8	3.8-36.2	D-210
cis-1,3-Dichloropropene	4.8-35.2	15.8	1.0-39.0	D-227
trans-1,3-Dichloropropene	10.0-30.0	10.4	7.6-32.4	17-183
Ethyl benzene	11.8-28.2	7.5	17.4-26.7	37-162
Methylene chloride	12.1-27.9	7.4	D-41.0	D-221
1,1,2,2-Tetrachloroethane	12.1-27.9	7.4	13.5-27.2	46-157
Tetrachloroethene	14.7-25.3	5.0	17.0-26.6	64-148
Toluene	14.9-25.1	4.8	16.6-26.7	47-150
1,1,1-Trichloroethane	15.0-25.0	4.6	13.7-30.1	52-162
1,1,2-Trichloroethane	14.2-25.8	5.5	14.3-27.1	52-150
Trichloroethene	13.3-26.7	6.6	18.6-27.6	71-157
Trichlorofluoromethane	9.6-30.4	10.0	8.9-31.5	17-181
Vinyl chloride	0.8-39.2	20.0	D-43.5	D-251

Q = Concentration measured in QC check sample, in μg/L (Section 7.5.3)

s = Standard deviation of four recovery measurements, in μg/L (Section 8.2.4)

$\bar{x}$ = Average recovery for four recovery measurements, in μg/L (Section 8.2.4)

P, P_s = Percent recovery measured, (Section 8.3.2, Section 8.4.2)

D = Detected; result must be greater than zero.

[a] Criteria were calculated assuming a QC check sample concentration of 20 μg/L.

Note: These criteria are based directly upon the method performance data in Table 6. Where necessary, the limits for recovery have been broadened to assure applicability of the limits to concentrations below those used to develop Table 6.

Method 625 – *Table 4*

Chromatographic Conditions, Method Detection Limits, and Characteristic Masses for Base/Neutral Extractables

Parameter	Reten-tion Time (min)	Method Detection Limit (µg/L)	Electron Impact Pri-mary	Electron Impact Second-ary	Electron Impact Second-ary	Chemical Ionization Meth-ane	Chemical Ionization Meth-ane	Chemical Ionization Meth-ane
1,3-Dichlorobenzene	7.4	1.9	146	148	113	146	148	150
1,4-Dichlorobenzene	7.8	4.4	146	148	113	146	148	150
Hexachloroethane	8.4	1.6	117	201	199	199	201	203
Bis(2-chloroethyl)ether	8.4	5.7	93	63	95	63	107	109
1,2-Dichlorobenzene	8.4	1.9	146	148	113	146	148	150
Bis(2-chloroisopropyl)ether[a]	9.3	5.7	45	77	79	77	135	137
N-Nitrosodi-n-propylamine			130	42	101			
Nitrobenzene	11.1	1.9	77	123	65	124	152	164
Hexachlorobutadiene	11.4	0.9	225	223	227	223	225	227
1,2,4-Trichlorobenzene	11.6	1.9	180	182	145	181	183	209
Isophorone	11.9	2.2	82	95	138	139	167	178
Naphthalene	12.1	1.6	128	129	127	129	157	169
Bis(2-chloroethoxy)methane	12.2	5.3	93	95	123	65	107	137
Hexachlorocyclopentadiene[a]	13.9		237	235	272	235	237	239
2-Chloronaphthalene	15.9	1.9	162	164	127	163	191	203
Acenaphthylene	17.4	3.5	152	151	153	152	153	181
Acenaphthene	17.8	1.9	154	153	152	154	155	183
Dimethyl phthalate	18.3	1.6	163	194	164	151	163	164
2,6-Dinitrotoluene	18.7	1.9	165	89	121	183	211	223
Fluorene	19.5	1.9	166	165	167	166	167	195
4-Chlorophenyl phenyl ether	19.5	4.2	204	206	141			
2,4-Dinitrotoluene	19.8	5.7	165	63	182	183	211	223
Diethyl phthalate	20.1	1.9	149	177	150	177	223	251
N-Nitrosodiphenylamine[b]	20.5	1.9	169	168	167	169	170	198
Hexachlorobenzene	21.0	1.9	284	142	249	284	286	288
β-BHC[b]	21.1		183	181	109			
4-Bromophenyl phenyl ether	21.2	1.9	248	250	141	249	251	277
δ-BHC[b]	22.4		183	181	109			
Phenanthrene	22.8	5.4	178	179	176	178	179	207
Anthracene	22.8	1.9	178	179	176	178	179	207
β-BHC	23.4	4.2	181	183	109			
Heptachlor	23.4	1.9	100	272	274			
δ-BHC	23.7	3.1	183	109	181			
Aldrin	24.0	1.9	66	263	220			
Dibutyl phthalate	24.7	2.5	149	150	104	149	205	279
Heptachlor epoxide	25.6	2.2	353	355	351			

Method 625 – *Table 4*

Chromatographic Conditions, Method Detection Limits, and Characteristic Masses for Base/Neutral Extractables (Continued)

Parameter	Retention Time (min)	Method Detection Limit (µg/L)	Electron Impact Primary	Electron Impact Secondary	Electron Impact Secondary	Chemical Ionization Methane	Chemical Ionization Methane	Chemical Ionization Methane
Endosulfan I[b]	26.4		237	339	341			
Fluoranthene	26.5	2.2	202	101	100	203	231	243
Dieldrin	27.2	2.5	79	263	279			
4,4'-DDE	27.2	5.6	246	248	176			
Pyrene	27.3	1.9	202	101	100	203	231	243
Endrin[b]	27.9		81	263	82			
Endosulfan II[b]	28.6		237	339	341			
4,4'-DDD	28.6	2.8	235	237	165			
Benzidine[b]	28.8	44	184	92	185	185	213	225
4,4'-DDT	29.3	4.7	235	237	165			
Endosulfan sulfate	29.8	5.6	272	387	422			
Endrin aldehyde			67	345	250			
Butyl benzyl phthalate	29.9	2.5	149	91	206	149	299	327
Bis(2-ethylhexyl)phthalate	30.6	2.5	149	167	279	149		
Chrysene	31.5	2.5	228	226	229	228	229	257
Benzo(a)anthracene	31.5	7.8	228	229	226	228	229	257
3,3'-Dichlorobenzidine	32.2	16.5	252	254	126			
Di-n-octyl phthalate	32.5	2.5	149					
Benzo(b)fluoranthene	34.9	4.8	252	253	125	252	253	281
Benzo(k)fluoranthene	34.9	2.5	252	253	125	252	253	281
Benzo(a)pyrene	36.4	2.5	252	253	125	252	253	281
Indeno(1,2,3-cd)pyrene	42.7	3.7	276	138	277	276	277	305
Dibenzo(a,h)anthracene	43.2	2.5	278	139	279	278	279	307
Benzo(ghi)perylene	45.1	4.1	276	138	277	276	277	305
N-Nitrosodimethylamine[b]			42	74	44			
Chlordane[c]	19-30		373	375	377			
Toxaphene[c]	25-34		159	231	233			
PCB-1016[c]	18-30		224	260	294			
PCB-1221[c]	15-30	30	190	224	260			
PCB-1232[c]	15-32		190	224	260			
PCB-1242[c]	15-32		224	260	294			
PCB-1248[c]	12-34		294	330	262			
PCB-1254[c]	22-34	36	294	330	362			
PCB-1260[c]	23-32		330	362	394			

[a] The proper chemical name is 2,2'-oxybis(1-chloropropane).
[b] See Section 1.2.
[c] These compounds are mixtures of various isomers. Column conditions: Supelcoport (100/120 mesh) coated with 3% SP-2250 packed in a 1.8 m long X 2 mm ID glass column with helium carrier gas at 30 mL/min. flow rate. Column temperature held isothermal at 50 °C for 4 min., then programmed at 8 °C/min. to 270 °C and held for 30 min.

Method 625 – *Table 5*

Chromatographic Conditions, Method Detection Limits, and Characteristic Masses for Acid Extractables

Parameter	Reten-tion Time (min)	Method Detection Limit (μg/L)	Characteristic Masses					
			Electron Impact			Chemical Ionization		
			Pri-mary	Second-ary	Second-ary	Meth-ane	Meth-ane	Meth-ane
2-Chlorophenol	5.9	3.3	128	64	130	129	131	157
2-Nitrophenol	6.5	3.6	139	65	109	140	168	122
Phenol	8.0	1.5	94	65	66	95	123	135
2,4-Dimethylphenol	9.4	2.7	122	107	121	123	151	163
2,4-Dichlorophenol	9.8	2.7	162	164	98	163	165	167
2,4,6-Trichlorophenol	11.8	2.7	196	198	200	197	199	201
4-Chloro-3-methylphenol	13.2	3.0	142	107	144	143	171	183
2,4-Dinitrophenol	15.9	42	184	63	154	185	213	225
2-Methyl-4,6-dinitrophenol	16.2	24	198	182	77	199	227	239
Pentachlorophenol	17.5	3.6	266	264	268	267	265	269
4-Nitrophenol	20.3	2.4	65	139	109	140	168	122

Column conditions: Supelcoport (100/120 mesh) coated with 1% SP-1240DA packed in a 1.8 m long X 2 mm ID glass column with helium carrier gas at 30 mL/min. flow rate. Column temperature held isothermal at 70 °C for 2 min. then programmed at 8 °C/min. to 200 °C.

Method 625 – *Table 6*

QC Acceptance Criteria

Parameter	Test conc. (μg/L)	Limit for s (μg/L)	Range for $\bar{x}$ (μg/L)	Range for P, P$_s$ (%)
Acenaphthene	100	27.6	60.1-132.3	47-145
Acenaphthylene	100	40.2	53.5-126.0	33-145
Aldrin	100	39.0	7.2-152.2	D-166
Anthracene	100	32.0	43.4-118.0	27-133
Benzo(a)anthracene	100	27.6	41.8-133.0	33-143
Benzo(b)fluoranthene	100	38.8	42.0-140.4	24-159
Benzo(k)fluoranthene	100	32.3	25.2-145.7	11-162
Benzo(a)pyrene	100	39.0	31.7-148.0	17-163
Benzo(ghi)perylene	100	58.9	D-195.0	D-219
Benzyl butyl phthalate	100	23.4	D-139.9	D-152
β-BHC	100	31.5	41.5-130.6	24-149
δ-BHC	100	21.6	D-100.0	D-110
Bis(2-chloroethyl)ether	100	55.0	42.9-126.0	12-158
Bis(2-chloroethoxy)methane	100	34.5	49.2-164.7	33-184
Bis(2-chloroisopropyl)ether[a]	100	46.3	62.8-138.6	36-166
Bis(2-ethylhexyl)phthalate	100	41.1	28.9-136.8	8-158
4-Bromophenyl phenyl ether	100	23.0	64.9-114.4	53-127
2-Chloronaphthalene	100	13.0	64.5-113.5	60-118
4-Chlorophenyl phenyl ether	100	33.4	38.4-144.7	25-158
Chrysene	100	48.3	44.1-139.9	17-168
4,4'-DDD	100	31.0	D-134.5	D-145
4,4'-DDE	100	32.0	19.2-119.7	4-136
4,4'-DDT	100	61.6	D-170.6	D-203
Dibenzo(a,h)anthracene	100	70.0	D-199.7	D-227
Di-n-butyl phthalate	100	16.7	8.4-111.0	1-118
1,2-Dichlorobenzene	100	30.9	48.6-112.0	32-129
1,3-Dichlorobenzene	100	41.7	16.7-153.9	D-172
1,4-Dichlorobenzene	100	32.1	37.3-105.7	20-124
3,3'-Dichlorobenzidine	100	71.4	8.2-212.5	D-262
Dieldrin	100	30.7	44.3-119.3	29-136
Diethyl phthalate	100	26.5	D-100.0	D-114
Dimethyl phthalate	100	23.2	D-100.0	D-112
2,4-Dinitrotoluene	100	21.8	47.5-126.9	39-139
2,6-Dinitrotoluene	100	29.6	68.1-136.7	50-158
Di-n-octyl phthalate	100	31.4	18.6-131.8	4-146
Endosulfan sulfate	100	16.7	D-103.5	D-107
Endrin aldehyde	100	32.5	D-188.8	D-209

Method 625 – *Table 6*

QC Acceptance Criteria (Continued)

Parameter	Test conc. (µg/L)	Limit for s (µg/L)	Range for x̄ (µg/L)	Range for P, Pₛ (%)
Fluoranthene	100	32.8	42.9-121.3	26-137
Fluorene	100	20.7	71.6-108.4	59-121
Heptachlor	100	37.2	D-172.2	D-192
Heptachlor epoxide	100	54.7	70.9-109.4	26-155
Hexachlorobenzene	100	24.9	7.8-141.5	D-152
Hexachlorobutadiene	100	26.3	37.8-102.2	24-116
Hexachloroethane	100	24.5	55.2-100.0	40-113
Indeno(1,2,3-cd)pyrene	100	44.6	D-150.9	D-171
Isophorone	100	63.3	46.6-180.2	21-196
Naphthalene	100	30.1	35.6-119.6	21-133
Nitrobenzene	100	39.3	54.3-157.6	35-180
N-Nitrosodi-n-propylamine	100	55.4	13.6-197.9	D-230
PCB-1260	100	54.2	19.3-121.0	D-164
Phenanthrene	100	20.6	65.2-108.7	54-120
Pyrene	100	25.2	69.6-100.0	52-115
1,2,4-Trichlorobenzene	100	28.1	57.3-129.2	44-142
4-Chloro-3-methylphenol	100	37.2	40.8-127.9	22-147
2-Chlorophenol	100	28.7	36.2-120.4	23-134
2,4-Dichlorophenol	100	26.4	52.5-121.7	39-135
2,4-Dimethylphenol	100	26.1	41.8-109.0	32-119
2,4-Dinitrophenol	100	49.8	D-172.9	D-191
2-Methyl-4,6-dinitrophenol	100	93.2	53.0-100.0	D-181
2-Nitrophenol	100	35.2	45.0-166.7	29-182
4-Nitrophenol	100	47.2	13.0-106.5	D-132
Pentachlorophenol	100	48.9	38.1-151.8	14-176
Phenol	100	22.6	16.6-100.0	5-112
2,4,6-Trichlorophenol	100	31.7	52.4-129.2	37-144

s   =   Standard deviation of four recovery measurements, in µg/L. (Section 8.2.4)

x̄   =   Average recovery for four recovery measurements, in µg/L. (Section 8.2.4)
P, Pₛ =   Percent recovery measured. (Section 8.3.2, Section 8.4.2)
D   =   Detected; result must be greater than zero.
Note: These criteria are based directly upon the method performance data in Table 7. Where necessary, the limits for recovery have been broadened to assure applicability of the limits to concentrations below those used to develop Table 7.

(a) The proper chemical name is 2,2'-oxybis(1-chloropropane).

Method 625 – *Table 8*

Suggested Internal and Surrogate Standards

Base/neutral fraction	Acid fraction
Aniline-d_5	2-Fluorophenol
Anthracene-d_{10}	Pentafluorophenol
Benzo(a)anthracene-d_{12}	Phenol-d_5
4,4'-Dibromobiphenyl	2-Perfluoromethyl phenol
4,4'-Dibromoctafluorobiphenyl	
Decafluorobiphenyl	
2,2'-Difluorobiphenyl	
4-Fluoroaniline	
1-Fluoronaphthalene	
2-Fluoronaphthalene	
Naphthalene-d_8	
Nitrobenzene-d_5	
2,3,4,5,6-Pentafluorobiphenyl	
Phenanthrene-d_{10}	
Pyridine-d_5	

Method 625 – *Table 9*

DFTPP KEY MASSES AND ION ABUNDANCE CRITERIA

Mass	m/z Abundance criteria
51	30-60 percent of mass 198
68	Less than 2 percent of mass 69
70	Less than 2 percent of mass 69
127	40-60 percent of mass 198
197	Less than 1 percent of mass 198
198	Base peak, 100 percent relative abundance
199	5-9 percent of mass 198
275	10-30 percent of mass 198
365	Greater than 1 percent of mass 198
441	Present but less than mass 443
442	Greater than 40 percent of mass 198
443	17-23 percent of mass 442

Method 3111 – *Table 3111:I*

Atomic Absorption Concentration Ranges with Direct Aspiration Atomic Absorption

Element	Wavelength nm	Flame Gases*	Instrument Detection Limit mg/L	Sensitivity mg/L	Optimum Concentration Range mg/L
Ag	328.1	A-Ac	0.01	0.06	0.1-4
Al	309.3	N-Ac	0.1	1	5-100
Au	242.8	A-Ac	0.01	0.25	0.5-20
Ba	553.6	N-Ac	0.03	0.4	1-20
Be	234.9	N-Ac	0.005	0.03	0.05-2
Bi	223.1	A-Ac	0.06	0.4	1-50
Ca	422.7	A-Ac	0.003	0.08	0.2-20
Cd	228.8	A-Ac	0.002	0.025	0.05-2
Co	240.7	A-Ac	0.03	0.2	0.5-10
Cr	357.9	A-Ac	0.02	0.1	0.2-10
Cs	852.1	A-Ac	0.02	0.3	0.5-15
Cu	324.7	A-Ac	0.01	0.1	0.2-10
Fe	248.3	A-Ac	0.02	0.12	0.3-10
Ir	264.0	A-Ac	0.6	8	-
K	766.5	A-Ac	0.005	0.04	0.1-2
Li	670.8	A-Ac	0.002	0.04	0.1-2
Mg	285.2	A-Ac	0.0005	0.007	0.02-2
Mn	279.5	A-Ac	0.01	0.05	0.1-10
Mo	313.3	N-Ac	0.1	0.5	1-20
Na	589.0	A-Ac	0.002	0.015	0.03-1
Ni	232.0	A-Ac	0.02	0.15	0.3-10
Os	290.9	N-Ac	0.08	1	-
Pb**	283.3	A-Ac	0.05	0.5	1-20
Pt	265.9	A-Ac	0.1	2.0	5-75
Rh	343.5	A-Ac	0.5	0.3	-
Ru	349.9	A-Ac	0.07	0.5	-
Sb	217.6	A-Ac	0.07	0.5	1-40
Si	251.6	N-Ac	0.3	2	5-150
Sn	224.6	A-Ac	0.8	4	10-200
Sr	460.7	A-Ac	0.03	0.15	0.3-5
Ti	365.3	N-Ac	0.3	2	5-100
V	318.4	N-Ac	0.2	1.5	2-100
Zn	213.9	A-Ac	0.005	0.02	0.05-2

*A-Ac = air-acetylene; N-Ac = nitrous oxide-acetylene

**The more sensitive 217.0 nm wavelength is recommended for instruments with background correction capabilities.

Method 3113 – *Table 3113:II*

Detection Levels and Concentration Ranges for Electrothermal Atomization Atomic Absorption Spectrometry

Element	Wavelength nm	Estimated Detection Limit µg/L	Optimum Concentration Range µg/L
Al	309.3	3	20-200
Sb	217.6	3	20-300
As*	193.7	1	5-100
Ba**	553.6	2	10-200
Be	234.9	0.2	1-30
Cd	228.8	0.1	0.5-10
Cr	357.9	2	5-100
Co	240.7	1	5-100
Cu	324.7	1	5-100
Fe	248.3	1	5-100
Pb***	283.3	1	5-100
Mn	279.5	0.2	1-30
Mo**	313.3	1	3-60
Ni**	232.0	1	5-100
Se*	196.0	2	5-100
Ag	328.1	0.2	1-25
Sn	224.6	5	20-300

*Gas interrupt utilized

**Pyrolytic graphite tubes utilized

***The more sensitive 217.0-nm wavelength is recommended for instruments with background correction capabilities.π

Method 3120 – *Table 3120:I*

Suggested Wavelengths, Estimated Detection Limits, Alternate Wavelengths, Calibration Concentrations, and Upper Limits

Element	Suggested Wavelength nm	Estimated Detection Limit µg/L	Alternate Wavelength* nm	Calibration Concentration mg/L	Upper Limit Concentration mg/L
Aluminum	308.22	40	237.32	10.0	100
Antimony	206.83	30	217.58	10.0	100
Arsenic	193.70	50	189.04**	10.0	100
Barium	455.40	2	493.41	1.0	50
Beryllium	313.04	0.3	234.86	1.0	10
Boron	249.77	5	249.68	1.0	50
Cadmium	226.50	4	214.44	2.0	50
Calcium	317.93	10	315.89	10.0	100
Chromium	267.72	7	206.15	5.0	50
Cobalt	228.62	7	230.79	2.0	50
Copper	324.75	6	219.96	1.0	50
Iron	259.94	7	238.20	10.0	100
Lead	220.35	40	217.00	10.0	100
Lithium	670.78	4***	-	5.0	100
Magnesium	279.08	30	279.55	10.0	100
Manganese	257.61	2	294.92	2.0	50
Molybdenum	202.03	8	203.84	10.0	100
Nickel	231.60	15	221.65	2.0	50
Potassium	766.49	100***	769.90	10.0	100
Selenium	196.03	75	203.99	5.0	100
Silica (SiO_2)	212.41	20	251.61	21.4	100
Silver	328.07	7	338.29	2.0	50
Sodium	589.00	30***	589.59	10.0	100
Strontium	407.77	0.5	421.55	1.0	50
Thallium	190.86**	40	377.57	10.0	100
Vanadium	292.40	8	-	1.0	50
Zinc	213.86	2	206.20	5.0	100

*Other wavelengths may be substituted if they provide the needed sensitivity and are corrected for spectral interference.
**Available with vacuum or inert gas purged optical path
***Sensitive to operating conditions

Method 6210 – *Table 6210:I*

Chromatographic Conditions and Method Detection Limits (MDL)

Compound	Retention Time min	Method Detection Limit µg/L
Chloromethane	2.3	nd
Bromomethane	3.1	nd
Vinyl chloride	3.8	nd
Chloroethane	4.6	nd
Methylene chloride	6.4	2.8
Trichlorofluoromethane	8.3	nd
1,1-Dichloroethene	9.0	2.8
1,1-Dichloroethane	10.1	4.7
trans-1,2-Dichloroethene	10.8	1.6
Chloroform	11.4	1.6
1,2-Dichloroethane	12.1	2.8
1,1,1-Trichloroethane	13.4	3.8
Carbon tetrachloride	13.7	2.8
Bromodichloromethane	14.3	2.2
1,2-Dichloropropane	15.7	6.0
cis-1,3-Dichloropropene	15.9	5.0
Trichloroethene	16.5	1.9
Benzene	17.0	4.4
Dibromochloromethane	17.1	3.1
1,1,2-Trichloroethane	17.2	5.0
trans-1,3-Dichloropropene	17.2	nd
2-Chloroethylvinyl ether	18.6	nd
Bromoform	19.8	4.7
1,1,2,2-Tetrachloroethane	22.1	6.9
Tetrachloroethene	22.2	4.1
Toluene	23.5	6.0
Chlorobenzene	24.6	6.0
Ethyl benzene	26.4	7.2
1,3-Dichlorobenzene	33.9	nd
1,2-Dichlorobenzene	35.0	nd
1,4-Dichlorobenzene	35.4	nd

Column conditions: Carbopack B (60/80 mesh) coated with 1% SP-1000 packed in a 1.8 m by 3 mm ID glass column with helium carrier gas at 30 mL/min flow rate. Column temperature held at 45 °C for 3 min, then programmed at 8 °C/min to 220 °C and held for 15 min.

nd = not determined

Method 6210 – *Table 6210:II*

BFB KEY m/z Abundance Criteria

Mass	m/z Abundance Criteria
50	15 to 40% of mass 95
75	30 to 60% of mass 95
95	Base peak, 100% relative abundance
96	5 to 9% of mass 95
173	<2% of mass 174
174	>50% of mass 95
175	5 to 9% of mass 174
176	>95% but <101% of mass 174
177	5 to 9% of mass 176

Method 6210 – *Table 6210:III*

Suggested Surrogate and Internal Standards

Compound	Retention Time* min	Primary m/z	Secondary Masses
Benzene-d$_6$	17.0	84	
4-Bromofluorobenzene	28.3	95	174, 176
1,2-Dichloroethane-d$_4$	12.1	102	
1,4-Difluorobenzene	19.6	114	63, 88
Ethylbenzene-d$_5$	26.4	111	
Ethylbenzene-d$_{10}$	26.4	98	
Fluorobenzene	18.4	96	70
Pentafluorobenzene	23.5	168	
Bromochloromethane	9.3	128	49, 130, 51
2-Bromo-1-chloropropane	19.2	77	79, 156
1,4-Dichlorobutane	25.8	55	90, 92

*For chromatographic conditions, see Table 6210:I. (page A-68)

Method 6210 – Table 6210:IV

Calibration and QC Acceptance Criteria*

Compound	Range for Q µg/L	Limit for s µg/L	Range for x̄ g/L	Range for P, P_s %
Benzene	12.8-27.2	6.9	15.2-26.0	37-151
Bromodichloromethane	13.1-26.9	6.4	10.1-28.0	35-155
Bromoform	14.2-25.8	5.4	11.4-31.1	45-169
Bromomethane	2.8-37.2	17.9	D-41.2	D-242
Carbon tetrachloride	14.6-25.4	5.2	17.2-23.5	70-140
Chlorobenzene	13.2-26.8	6.3	16.4-27.4	37-160
Chloroethane	7.6-32.4	11.4	8.4-40.4	14-230
2-Chloroethylvinyl ether	D-44.8	25.9	D-50.4	D-305
Chloroform	13.5-26.5	6.1	13.7-24.2	51-138
Chloromethane	D-40.8	19.8	D-45.9	D-273
Dibromochloromethane	13.5-26.5	6.1	13.8-26.6	53-149
1,2-Dichlorobenzene	12.6-27.4	7.1	11.8-34.7	18-190
1,3-Dichlorobenzene	14.6-25.4	5.5	17.0-28.8	59-156
1,4-Dichlorobenzene	12.6-27.4	7.1	11.8-34.7	18-190
1,1-Dichloroethane	14.5-25.5	5.1	14.2-28.5	59-155
1,2-Dichloroethane	13.6-26.4	6.0	14.3-27.4	49-155
1,1-Dichloroethene	10.1-29.9	9.1	3.7-42.3	D-234
trans-1,2-Dichloroethene	13.9-26.1	5.7	13.6-28.5	54-156
1,2-Dichloropropane	6.8-33.2	13.8	3.8-36.2	D-210
cis-1,3-Dichloropropene	4.8-35.2	15.8	1.0-39.0	D-227
trans-1,3-Dichloropropene	10.0-30.0	10.4	7.6-32.4	17-183
Ethyl benzene	11.8-28.2	7.5	17.4-26.7	37-162
Methylene chloride	12.1-27.9	7.4	D-41.0	D-221
1,1,2,2-Tetrachloroethane	12.1-27.9	7.4	13.5-27.2	46-157
Tetrachloroethene	14.7-25.3	5.5	17.0-26.6	64-148
Toluene	14.9-25.1	4.8	16.6-26.7	47-150
1,1,1-Trichloroethane	15.0-25.0	4.6	13.7-30.1	52-162
1,1,2-Trichloroethane	14.2-25.8	5.5	14.3-27.1	52-150
Trichloroethene	13.3-26.7	6.6	18.6-27.6	71-157
Trichlorofluoromethane	9.6-30.4	10.0	8.9-31.5	17-181
Vinyl chloride	0.8-39.2	20.0	D-43.5	D-251

*Q = concentration measured in QC check sample

s = standard deviation of four recovery measurements

x̄ = average recovery of four recovery measurements

P, P_s = percent recovery measured

D = Detected; result must be greater than zero.

Criteria calculated assuming a QC check sample concentration of 20 µg/L

Note: These criteria are based directly on the method performance data in Table 6210:VI. Where necessary, the limits for recovery were broadened to assure applicability of the limits to concentrations below those used to develop Table 6210:VI.

Method 6220 – *Table 6220:I*

Chromatographic Conditions and Method Detection Limits

Compound	Retention Time (min)		Method Detection
	Column 1	Column 2	Limit µg/L
Benzene	3.33	2.75	0.2
Toluene	5.75	4.25	0.2
Ethylbenzene	8.25	6.25	0.2
Chlorobenzene	9.17	8.02	0.2
1,4-Dichlorobenzene	16.8	16.2	0.3
1,3-Dichlorobenzene	18.2	15.0	0.4
1,2-Dichlorobenzene	25.9	19.4	0.4

Column 1 conditions: Supelcoport (100/120 mesh) coated with 5% SP-1200/1, 75% Bentone-34 packed in a 1.8 m x 2.2 mm ID stainless steel column with helium carrier gas at 36 mL/min flow rate. Column temperature held at 50 °C for 2 min, then programmed at 6 °C/min to 90°C for a final hold.

Column 2 conditions: Chromosorb W-AW (60/80 mesh) coated with 5% 1,2,3-tris(2-cyanoethoxy)propane packed in a 1.8 m x 2.2 mm ID stainless steel column with helium carrier gas at 30 mL/min flow rate. Column temperature held at 40 °C for 2 min, then programmed at 2 °C/min to 100 °C for a final hold

Method 6220 – *Table 6220:II*

Calibration and QC Acceptance Criteria*

Compound	Range for Q µg/L	Limit for s µg/L	Range for x̄ µg/L	Range for P, P_s %
Benzene	15.4-24.6	4.1	10.0-27.9	39-150
Chlorobenzene	16.1-23.9	3.5	12.7-25.4	55-135
1,2-Dichlorobenzene	13.6-26.4	5.8	10.6-27.6	37-154
1,3-Dichlorobenzene	14.5-25.5	5	12.8-25.5	50-141
1,4-Dichlorobenzene	13.9-26.1	5.5	11.6-25.5	42-143
Ethylbenzene	12.6-27.4	6.7	10.0-28.2	32-160
Toluene	15.5-24.5	4	11.2-27.7	46-148

*Q   =   concentration measured in QC check sample
s   =   standard deviation of four recovery measurements
x̄   =   average recovery for four recovery measurements
P, P_s =   recovery measured

Criteria calculated assuming a QC check sample concentration of 20 µg/L

NOTE: These criteria are based directly on the method performance data in Table 6220:III. Where necessary, the limits for recovery were broadened to assure applicability of the limits to concentrations below those used to develop Table 6220:III.

Method 6230 – *Table 6230:I*

Chromatographic Conditions and Method Detection Limits (MDL)

Compound	Retention Time (min)		Method Detection Limit µg/L
	Column 1	Column 2	
Chloromethane	1.50	5.28	0.08
Bromomethane	2.17	7.05	1.18
Dichlorodifluoromethane	2.62	nd	1.81
Vinyl chloride	2.67	5.28	0.18
Chloroethane	3.33	8.68	0.52
Methylene chloride	5.25	10.1	0.25
Trichlorofluoromethane	7.18	nd	nd
1,1-Dichloroethene	7.93	7.72	0.13
1,1-Dichloroethane	9.30	12.6	0.07
trans-1,2-Dichloroethene	10.1	9.38	0.10
Chloroform	10.7	12.1	0.05
1,2-Dichloroethane	11.4	15.4	0.03
1,1,1-Trichloroethane	12.6	13.1	0.03
Carbon tetrachloride	13.0	14.4	0.12
Bromodichloromethane	13.7	14.6	0.10
1,2-Dichloropropane	14.9	16.6	0.04
cis-1,3-Dichloropropene	15.2	16.6	0.34
Trichloroethene	15.8	13.1	0.12
Dibromochloromethane	16.5	16.6	0.09
1,1,2-Trichloroethane	16.5	18.1	0.02
trans-1,3-Dichloropropene	16.5	18	0.20
2-Chloroethylvinyl ether	18.0	nd	0.13
Bromoform	19.2	19.2	0.20
1,1,2,2-Tetrachloroethane	21.6	nd	0.03
Tetrachloroethene	21.7	15	0.03
Chlorobenzene	24.2	18.8	0.25
1,3-Dichlorobenzene	34.0	22.4	0.32
1,2-Dichlorobenzene	34.9	23.5	0.15
1,4-Dichlorobenzene	35.4	22.3	0.24

Column 1 conditions: Carbopack B (60/80 mesh) coated with 1% SP-1000 packed in a 2.4 m x 3 mm ID stainless steel or glass column with helium carrier gas at 40 mL/min flow rate. Column temperature held at 45 °C for 3 min, then programmed at 8 °C/min to 220 °C and held for 15 min.

Column 2 conditions: Porisil-C (100/120 mesh) coated with n-octane packed in a 1.8 m x 3 mm ID stainless steel or glass column with helium carrier gas at 40 mL/min flow rate. Column temperature held at 50 °C for 3 min, then programmed at 6 °C/min to 170 °C and held for 4 min.

nd = not determined

Method 6230 – *Table 6230:II*

Calibration and QC Acceptance Criteria*

Compounds	Range for Q µg/L	Limit for s µg/L	Range for x̄ µg/L	Range for P,P$_s$ %
Bromodichloromethane	15.2-24.8	4.3	10.7-32.0	42-172
Bromoform	14.7-25.3	4.7	5.0-29.3	13-159
Bromomethane	11.7-28.3	7.6	3.4-24.5	D-144
Carbon tetrachloride	13.7-26.3	5.6	11.8-25.3	43-143
Chlorobenzene	14.4-25.6	5.0	10.2-27.4	38-150
Chloroethane	15.4-24.6	4.4	11.3-25.2	46-137
2-Chloroethylvinyl ether	12.0-28.0	8.3	4.5-35.5	14-186
Chloroform	15.0-25.0	4.5	12.4-24.0	49-133
Chloromethane	11.9-28.1	7.4	D-34.9	D-193
Dibromochloromethane	13.1-26.9	6.3	7.9-35.1	24-191
1,2-Dichlorobenzene	14.0-26.0	5.5	1.7-38.9	D-208
1,3-Dichlorobenzene	9.9-30.1	9.1	6.2-32.6	7-187
1,4-Dichlorobenzene	13.9-26.1	5.5	11.5-25.5	42-143
1,1-Dichloroethane	16.8-23.2	3.2	11.2-24.6	47-132
1,2-Dichloroethane	14.3-25.7	5.2	13.0-26.5	51-147
1,1-Dichloroethene	12.6-27.4	6.6	10.2-27.3	28-167
trans-1,2-Dichloroethene	12.8-27.2	6.4	11.4-27.1	38-155
1,2-Dichloropropane	14.8-25.2	5.2	10.1-29.9	44-156
cis-1,3-Dichloropropene	12.8-27.2	7.3	6.2-33.8	22-178
trans-1,3-Dichloropropene	12.8-27.2	7.3	6.2-33.8	22-178
Methylene chloride	15.5-24.5	4.0	7.0-27.6	25-162
1,1,2,2-Tetrachloroethane	9.8-30.2	9.2	6.6-31.8	8-184
Tetrachloroethene	14.0-26.0	5.4	8.1-29.6	26-162
1,1,1-Trichloroethane	14.2-25.8	4.9	10.8-24.8	41-138
1,1,2-Trichloroethane	15.7-24.3	3.9	9.6-25.4	39-136
Trichloroethene	15.4-24.6	4.2	9.2-26.6	35-146
Trichlorofluoromethane	13.3-26.7	6.0	7.4-28.1	21-156
Vinyl chloride	13.7-26.3	5.7	8.2-29.9	28-163

*Q   = concentration measured in QC check sample

s   = standard deviation of four recovery measurements

x̄   = average recovery for four recovery measurements

P, P$_s$ = recovery measured

D   = detected; result must be greater than zero.

Criteria calculated assuming a QC check sample concentration of 20 ug/L

NOTE: These criteria are based directly on the method performance data in Table 6230:III. Where necessary, the limits for recovery were broadened to assure applicability of the limits to concentrations below those used to develop Table 6230:III.

Method 6230 – *Table 6230:III*

Method Bias and Precision as Functions of Concentration*

Compound	Bias, as Recovery, X' µg/L	Single-Analyst Precision, s' µg/L	Overall Precision, S' µg/L
Bromodichloromethane	1.12C-1.02	$0.11\bar{x}$ +0.04	$0.20\bar{x}$ +1.00
Bromoform	0.96C-2.05	$0.12\bar{x}$ +0.58	$0.21\bar{x}$ +2.41
Bromomethane	0.76C-1.27	$0.28\bar{x}$ +0.27	$0.36\bar{x}$ +0.94
Carbon tetrachloride	0.98C-1.04	$0.15\bar{x}$ +0.38	$0.20\bar{x}$ +0.39
Chlorobenzene	1.00C-1.23	$0.15\bar{x}$ -0.02	$0.18\bar{x}$ +1.21
Chloroethane	0.99C-1.53	$0.14\bar{x}$ -0.13	$0.17\bar{x}$ +0.63
2-Chloroethylvinyl ether**	1.00C	$0.20\bar{x}$	$0.35\bar{x}$
Chloroform	0.93C-0.39	$0.13\bar{x}$ +0.15	$0.19\bar{x}$ -0.02
Chloromethane	0.77C+0.18	$0.28\bar{x}$ -0.31	$0.52\bar{x}$ +1.31
Dibromochloromethane	0.94C+2.72	$0.11\bar{x}$ +1.10	$0.24\bar{x}$ +1.68
1,2-Dichlorobenzene	0.93C+1.70	$0.20\bar{x}$ +0.97	$0.13\bar{x}$ +6.13
1,3-Dichlorobenzene	0.95C+0.43	$0.14\bar{x}$ +2.33	$0.26\bar{x}$ +2.34
1,4-Dichlorobenzene	0.93C-0.09	$0.15\bar{x}$ +0.29	$0.20\bar{x}$ +0.41
1,1-Dichloroethane	0.95C-1.08	$0.08\bar{x}$ +0.17	$0.14\bar{x}$ +0.94
1,2-Dichloroethane	1.04C-1.06	$0.11\bar{x}$ +0.70	$0.15\bar{x}$ +0.94
1,1-Dichloroethene	0.98C-0.87	$0.21\bar{x}$ -0.23	$0.29\bar{x}$ -0.40
trans-1,2-Dichloroethene	0.97C-0.16	$0.11\bar{x}$ +1.46	$0.17\bar{x}$ +1.46
1,2-Dichloropropane**	1.00C	$0.13\bar{x}$	$0.23\bar{x}$
cis-1,3-Dichloropropene**	1.00C	$0.18\bar{x}$	$0.32\bar{x}$
trans-1,3-Dichloropropene**	1.00C	$0.18\bar{x}$	$0.32\bar{x}$
Methylene chloride	0.91C-0.93	$0.11\bar{x}$ +0.33	$0.21\bar{x}$ +1.43
1,1,2,2-Tetrachloroethene	0.95C+0.19	$0.14\bar{x}$ +2.41	$0.23\bar{x}$ +2.79
Tetrachloroethene	0.94C+0.06	$0.14\bar{x}$ +0.38	$0.18\bar{x}$ +2.21
1,1,1-Trichloroethane	0.90C-0.16	$0.15\bar{x}$ +0.04	$0.20\bar{x}$ +0.37
1,1,2-Trichloroethane	0.86C+0.30	$0.13\bar{x}$ -0.14	$0.19\bar{x}$ +0.67
Trichloroethene	0.87C+0.48	$0.13\bar{x}$ -0.03	$0.23\bar{x}$ +0.30
Trichlorofluoromethane	0.89C-0.07	$0.15\bar{x}$ +0.67	$0.26\bar{x}$ +0.91
Vinyl chloride	0.97C-0.36	$0.13\bar{x}$ +0.65	$0.27\bar{x}$ +0.40

*X' = expected recovery for one or more measurements of a sample containing a concentration of C

s' = expected single-analyst standard deviation of measurements at an average concentration found of $\bar{x}$

S' = expected interlaboratory standard deviation of measurements at an average concentration found of $\bar{x}$

C = true value for the concentration, µg/L

$\bar{x}$ = average recovery found for measurements of samples containing a concentration of C, µg/L

**Estimates based on performance in a single laboratory

Method 6410 – *Table 6410:I*

Chromatographic Conditions, Method Detection Limits, and Characteristic Masses for Base/Neutral Extractables

Compound	Retention Time min	Method Detection Limit µg/L	Characteristic Masses					
			Electron Impact			Chemical Ionization		
			Primary	Secondary	Secondary	Methane	Methane	Methane
1,3-Dichlorobenzene	7.4	1.9	146	148	113	146	148	150
1,4-Dichlorobenzene	7.8	4.4	146	148	113	146	148	150
Hexachloroethane	8.4	1.6	117	201	199	199	201	203
bis(2-Chloroethyl) ether	8.4	5.7	93	63	95	63	107	109
1,2-Dichlorobenzene	8.4	1.9	146	148	113	146	148	150
bis(2-Chloroisopropyl) ether[a]	9.3	5.7	45	77	79	77	135	137
N-Nitrosodi-n-propylamine			130	42	101			
Nitrobenzene	11.1	1.9	77	123	65	124	152	164
Hexachlorobutadiene	11.4	0.9	225	223	227	223	225	227
1,2,4-Trichlorobenzene	11.6	1.9	180	182	145	181	183	209
Isophorone	11.9	2.2	82	95	138	139	167	178
Naphthalene	12.1	1.6	128	129	127	129	157	169
bis(2-Chloroethoxy) methane	12.2	5.3	93	95	123	65	107	137
Hexachlorocyclopentadiene[b]	13.9		237	235	272	235	237	239
2-Chloronaphthalene	15.9	1.9	162	164	127	163	191	203
Acenaphthylene	17.4	3.5	152	151	153	152	153	181
Acenaphthene	17.8	1.9	154	153	152	154	155	183
Dimethyl phthalate	18.3	1.6	163	194	164	151	163	164
2,6-Dinitrotoluene	18.7	1.9	165	89	121	183	211	223
Fluorene	19.5	1.9	166	165	167	166	167	195
4-Chlorophenyl phenyl ether	19.5	4.2	204	206	141			
2,4-Dinitrotoluene	19.8	5.7	165	63	182	183	211	223
Diethyl phthalate	20.1	1.9	149	177	150	177	223	251
N-Nitrosodiphenylamine[b]	20.5	1.9	169	168	167	169	170	198
Hexachlorobenzene	21.0	1.9	284	142	249	284	286	288
α-BHC[b]	21.1		183	181	109			
4-Bromophenyl phenyl ether	21.2	1.9	248	250	141	249	251	277
γ-BHC[b]	22.4		183	181	109			
Phenanthrene	22.8	5.4	178	179	176	178	179	207
Anthracene	22.8	1.9	178	179	176	178	179	207
β-BHC	23.4	4.2	181	183	109			
Heptachlor	23.4	1.9	100	272	274			

Method 6410 – *Table 6410:I*

Chromatographic Conditions, Method Detection Limits, and Characteristic Masses for Base/Neutral Extractables (Continued)

Compound	Retention Time min	Method Detection Limit µg/L	Characteristic Masses					
			Electron Impact			Chemical Ionization		
			Pri-mary	Second-ary	Second-ary	Meth-ane	Meth-ane	Meth-ane
δ-BHC	23.7	3.1	183	109	181			
Aldrin	24.0	1.9	66	263	220			
Dibutyl phthalate	24.7	2.5	149	150	104	149	205	279
Heptachlor epoxide	25.6	2.2	353	355	351			
Endosulfan I[b]	26.4		237	338	341			
Fluoranthene	26.5	2.2	202	101	100	203	231	243
Dieldrin	27.2	2.5	79	263	279			
4,4'-DDE	27.2	5.6	246	248	176			
Pyrene	27.3	1.9	202	101	100	203	231	243
Endrin[b]	27.9		81	263	82			
Endosulfan II[b]	28.6		237	339	341			
4,4'-DDD	28.6	2.8	235	237	165			
Benzidine[b]	28.8	44	184	92	185	185	213	225
4,4'-DDT	29.3	4.7	235	237	165			
Endosulfan sulfate	29.8	5.6	272	387	422			
Endrin aldehyde			67	345	250			
Butyl benzyl phthalate	29.9	2.5	149	91	206	149	299	327
bis(2-Ethylhexyl) phthalate	30.6	2.5	149	167	279	149		
Chrysene	31.5	2.5	228	226	229	228	229	257
Benzo(a)anthracene	31.5	7.8	228	229	226	228	229	257
3,3'-Dichlorobenzidine	32.2	16.5	252	254	126			
Di-n-octyl phthalate	32.5	2.5	149					
Benzo(b)fluoranthene	34.9	4.8	252	253	125	252	253	281
Benzo(k)fluoranthene	34.9	2.5	252	253	125	252	253	281
Benzo(a)pyrene	36.4	2.5	252	253	125	252	253	281
Indeno(1,2,3-cd)pyrene	42.7	3.7	276	138	277	276	277	305
Dibenzo(a,h)anthracene	43.2	2.5	278	139	279	278	279	307
Benzo(ghi)perylene	45.1	4.1	276	138	277	276	277	305
N-Nitrosodimethylamine[b]			42	74	44			
Chlordane[c]	19-30		373	375	377			
Toxaphene[c]	25-34		159	231	233			

Method 6410 – *Table 6410:I*

Chromatographic Conditions, Method Detection Limits, and Characteristic Masses for Base/Neutral Extractables (Continued)

Compound	Retention Time min	Method Detection Limit µg/L	Characteristic Masses					
			Electron Impact			Chemical Ionization		
			Pri-mary	Second-ary	Second-ary	Meth-ane	Meth-ane	Meth-ane
PCB 1016[c]	18-30		224	260	294			
PCB 1221[c]	15-30	30	190	224	260			
PCB 1232[c]	15-32		190	224	260			
PCB 1242[c]	15-32		224	260	294			
PCB 1248[c]	12-34		294	330	262			
PCB 1254[c]	22-34	36	294	330	362			
PCB 1260[c]	23-32		330	362	394			

[a]   The proper chemical name is 2,2'-oxybis(1-chloropropane).

[b]   See introductory section of text.

[c]   These compounds are mixtures of various isomers.

Column conditions: Supelcoport (100/120 mesh) coated with 3% SP-2250 packed in a 1.8 m long x 2 mm ID glass column with helium carrier gas at 30 mL/min flow rate. Column temperature held isothermal at 50 °C for 4 min, then programmed at 8 °C/min to 270 °C and held for 30 min.

Method 6410 – *Table 6410:II*

Chromatographic Conditions, Method Detection Limits, and Characteristic Masses for Acid Extractables

Compound	Retention Time min	Method Detection Limit µg/L	Characteristic Masses					
			Electron Impact			Chemical Ionization		
			Primary	Secondary	Secondary	Methane	Methane	Methane
2-Chlorophenol	5.9	3.3	128	64	130	129	131	157
2-Nitrophenol	6.5	3.6	139	65	109	140	168	122
Phenol	8.0	1.5	94	65	66	95	123	135
2,4-Dimethylphenol	9.4	2.7	122	107	121	123	151	163
2,4-Dichlorophenol	9.8	2.7	162	164	98	163	165	167
2,4,6-Trichlorophenol	11.8	2.7	196	198	200	197	199	201
4-Chloro-3-methylphenol	13.2	3.0	142	107	144	143	171	183
2,4-Dinitrophenol	15.9	42	184	63	154	185	213	225
2-Methyl-4,6-dinitrophenol	16.2	24	198	182	77	199	227	239
Pentachlorophenol	17.5	3.6	266	264	268	267	265	269
4-Nitrophenol	20.3	2.4	65	139	109	140	168	122

Column conditions: Supelcoport (100/120 mesh) coated with 1% SP-1240DA packed in a 1.8 m long x 2 mm ID glass column with helium carrier gas at 30 mL/min flow rate. Column temperature held isothermal at 70 °C for 2 min then programmed at 8 °C/min to 200 °C.

Method 6410 – *Table 6410:III*

DFTPP Key Masses and Abundance Criteria

Mass	m/z Abundance Criteria
51	30-60% of mass 198
68	Less than 2% of mass 69
70	Less than 2% of mass 69
127	40-60% of mass 198
197	Less than 1% of mass 198
198	Base peak, 100% relative abundance
199	5-9% of mass 198
275	10-30% of mass 198
365	Greater than 1% of mass 198
441	Present but less than mass 443
442	Greater than 40% of mass 198
443	17-23% of mass 442

Method 6410 – *Table 6410:IV*

Suggested Internal and Surrogate Standards

Base/Neutral Fraction	Acid Fraction
Aniline-d$_5$	2-Fluorophenol
Anthracene-d$_{10}$	Pentafluorophenol
Benzo(a)anthracene-d$_{12}$	Phenol-d$_5$
4,4'-Dibromobiphenyl	2-Perfluoromethyl phenol
4,4'-Dibromooctafluorobiphenyl	
Decafluorobiphenyl	
2,2'-Difluorobiphenyl	
4-Fluoroaniline	
1-Fluoronaphthylene	
2-Fluoronaphthylene	
Naphthalene-d$_8$	
Nitrobenzene-d$_5$	
2,3,4,5,6-Pentafluorobiphenyl	
Phenanthrene-d$_{10}$	
Pyridine-d$_5$	

Method 6410 – *Table 6410:V*

QC Acceptance Criteria*

Compound	Test Concentration μg/L	Limits for s μg/L	Range for $\bar{x}$ μg/L	Range for P,P$_s$ %
Acenaphthene	100	27.6	60.1-132.3	47-145
Acenaphthylene	100	40.2	53.5-126.0	33-145
Aldrin	100	39.0	7.2-152.2	D-166
Anthracene	100	32.0	43.4-118.0	27-133
Benzo(a)anthracene	100	27.6	41.8-133.0	33-143
Benzo(b)fluoranthene	100	38.8	42.0-140.4	24-159
Benzo(k)fluoranthene	100	32.3	25.2-145.7	11-162
Benzo(a)pyrene	100	39.0	31.7-148.0	17-163
Benzo(g,h,i)perylene	100	58.9	D-195.0	D-219
Benzyl butyl phthalate	100	23.4	D-139.9	D-152
δ-BHC	100	31.5	41.5-130.6	24-149
β-BHC	100	21.6	D-100.0	D-110
bis(2-Chloroethyl) ether	100	55.0	42.9-126.0	12-158
bis(2-Chloroethoxy) methane	100	34.5	49.2-164.7	33-184
bis(2-Chloroisopropyl) ether**	100	46.3	62.8-138.6	36-166
bis(2-Ethylhexyl) phthalate	100	41.1	28.9-136.8	8-158
4-Bromophenyl phenyl ether	100	23.0	64.9-114.4	53-127
2-Chloronapthalene	100	13.0	64.5-113.5	60-118
4-Chlorophenyl phenyl ether	100	33.4	38.4-144.7	25-158
Chrysene	100	48.3	44.1-139.9	17-168
4,4'-DDD	100	31.0	D-134.5	D-145
4,4'-DDE	100	32.0	19.2-119.7	4-136
4,4'-DDT	100	61.6	D-170.6	D-203
Dibenzo(a,h)anthracene	100	70.0	D-199.7	D-227
Di-n-butyl phthalate	100	16.7	8.4-111.0	1-118
1,2-Dichlorobenzene	100	30.9	48.6-112.0	32-129
1,3-Dichlorobenzene	100	41.7	16.7-153.9	D-172
1,4-Dichlorobenzene	100	32.1	37.3-105.7	20-124
3,3'-Dichlorobenzidine	100	71.4	8.2-212.5	D-262
Dieldrin	100	30.7	44.3-119.3	29-136
Diethyl phthalate	100	26.5	D-100.0	D-114
Dimethyl phthalate	100	23.2	D-100.0	D-112
2,4-Dinitrotoluene	100	21.8	47.5-126.9	39-139
2,6-Dinitrotoluene	100	29.6	68.1-136.7	50-158
Di-n-octylphthalate	100	31.4	18.6-131.8	4-146
Endosulfan sulfate	100	16.7	D-103.5	D-107
Endrin aldehyde	100	32.5	D-188.8	D-209

Method 6410 – *Table 6410:V*

QC Acceptance Criteria* (Continued)

Compound	Test Concentration μg/L	Limits for s μg/L	Range for x̄ μg/L	Range for P,P$_s$ %
Fluoranthene	100	32.8	42.9-121.3	26-137
Fluorene	100	20.7	71.6-108.4	59-121
Heptachlor	100	37.2	D-172.2	D-192
Heptachlor epoxide	100	54.7	70.9-109.4	26-155
Hexachlorobenzene	100	24.9	7.8-141.5	D-152
Hexachlorobutadiene	100	26.3	37.8-102.2	24-116
Hexachloroethane	100	24.5	55.2-100.0	40-113
Indeno(1,2,3-cd)pyrene	100	44.6	D-150.9	D-171
Isophorone	100	63.3	46.6-180.2	21-196
Naphthalene	100	30.1	35.6-119.6	21-133
Nitrobenzene	100	39.3	54.3-157.6	35-180
N-Nitrosodi-n-propylamine	100	55.4	13.6-197.9	D-230
PCB-1260	100	54.2	19.3-121.0	D-164
Phenanthrene	100	20.6	65.2-108.7	54-120
Pyrene	100	25.2	69.6-100.0	52-115
1,2,4-Trichlorobenzene	100	28.1	57.3-129.2	44-142
4-Chloro-3-methylphenol	100	37.2	40.8-127.9	22-147
2-Chlorophenol	100	28.7	36.2-120.4	23-134
2,4-Dichlorophenol	100	26.4	52.5-121.7	39-135
2,4-Dimethylphenol	100	26.1	41.8-109.0	32-119
2,4-Dinitrophenol	100	49.8	D-172.9	D-191
2-Methyl-4,6-dinitrophenol	100	93.2	53.0-100.0	D-181
2-Nitrophenol	100	35.2	45.0-166.7	29-182
4-Nitrophenol	100	47.2	13.0-106.5	D-132
Pentachlorophenol	100	48.9	38.1-151.8	14-176
Phenol	100	22.6	16.6-100.0	5-112
2,4,6-Trichlorophenol	100	31.7	52.4-129.2	37-144

*s   = standard deviation for four recovery measurements

x̄   = average recovery for four recovery measurements

P, P$_s$ = percent recovery measured

D   = detected; results must be greater than zero.

**The proper chemical name is 2,2'-oxybis(1-chloropropane).

NOTE: These criteria are based directly upon the method performance data in Table 6410:VI. Where necessary, the limits for recovery were broadened to assure applicability of the limits to concentrations below those used to develop Table 6410:VI.

Method 6630 – *Table 6630:III*

Chromatographic Conditions and Method Detection Limits[a]

Parameter	Retention time (min)		Method Detection Limit (μg/L)
	Col. 1	Col. 2	
α-BHC	1.35	1.82	0.003
γ-BHC (Lindane)	1.70	2.13	nd
β-BHC	1.90	1.97	nd
Heptachlor	2.00	3.35	0.003
δ-BHC	2.15	2.20	0.009
Aldrin	2.40	4.10	0.004
Heptachlor epoxide	3.50	5.00	0.083
Endosulfan I	4.50	6.20	0.014
4,4'-DDE	5.13	7.15	0.004
Dieldrin	5.45	7.23	0.002
Endrin	6.55	8.10	0.006
4,4'-DDD	7.83	9.08	0.011
Endosulfan II	8.00	8.28	0.004
4,4'-DDT	9.40	11.75	0.012
Endrin aldehyde	11.82	9.30	0.023
Endosulfan sulfate	14.22	10.70	0.066
Chlordane	mr	mr	0.014
Toxaphene	mr	mr	0.24
PCB-1016	mr	mr	nd
PCB-1221	mr	mr	nd
PCB-1232	mr	mr	nd
PCB-1242	mr	mr	0.065
PCB-1248	mr	mr	nd
PCB-1254	mr	mr	nd
PCB-1260	mr	mr	nd

Column 1 conditions: Supelcoport (100/120 mesh) coated with 1.5% SP-2250/1.95% SP-2401 packed in a 1.8 m long X 4 mm ID glass column with 5% methane/95% argon carrier gas at 60 mL/min flow rate. Column temperature held isothermal at 200 °C, except for PCB-1016 through PCB-1248, should be measured at 160 °C.

Column 2 conditions: Supelcoport (100/120 mesh) coated with 3% OV-1 packed in a 1.8 m long X 4 mm ID glass column with 5% methane/95% argon carrier gas at 60 mL/min flow rate. Column temperature held isothermal at 200 °C for the pesticides; at 140 °C for PCB-1221 and 1232; and at 170 °C for PCB-1016 and 1242 to 1268.

mr = Multiple peak response

nd = not determined

Method 6630 – *Table 6630:V*

QC Acceptance Criteria*

Parameter	Test conc. (µg/L)	Limit for s (µg/L)	Range for x̄ (µg/L)	Range for P, Ps (%)
Aldrin	2.0	0.042	1.08-2.24	42-122
α-BHC	2.0	0.48	0.98-2.44	37-134
β-BHC	2.0	0.64	0.78-2.60	17-147
δ-BHC	2.0	0.72	1.01-2.37	19-140
γ-BHC	2.0	0.46	0.86-2.32	32-127
Chlordane	50	10.0	27.6-54.3	45-119
4,4'-DDD	10	2.8	4.8-12.6	31-141
4,4'-DDE	2.0	0.55	1.08-2.60	30-145
4,4'-DDT	10	3.6	4.6-13.7	25-160
Dieldrin	2.0	0.76	1.15-2.49	36-146
Endosulfan I	2.0	0.49	1.14-2.82	45-153
Endosulfan II	10	6.1	2.2-17.1	D-202
Endosulfan sulfate	10	2.7	3.8-13.2	26-144
Endrin	10	3.7	5.1-12.6	30-147
Heptachlor	2.0	0.4	0.86-2.00	34-111
Heptachlor epoxide	2.0	0.41	1.13-2.63	37-142
Toxaphene	50	12.7	27.8-55.6	41-126
PCB-1016	50	10.0	30.5-51.5	50-114
PCB-1221	50	24.4	22.1-75.2	15-178
PCB-1232	50	17.9	14.0-98.5	10-215
PCB-1242	50	12.2	24.8-69.6	39-150
PCB-1248	50	15.9	29.0-70.2	38-158
PCB-1254	50	13.8	22.2-57.9	29-131
PCB-1260	50	10.4	18.7-54.9	8-127

s   =   Standard deviation of four recovery measurements, in µg/L (Section 8.2.4)
x̄   =   Average recovery for four recovery measurements, in µg/L (Section 8.2.4)
P, Ps =   Percent recovery measured (Section 8.3.2, Section 8.4.2)
D   =   Detected; result must be greater than zero.

Note: These criteria are based directly upon the method performance data in 6630:IV. Where necessary, the limits for recovery have been broadened to assure applicability of the limits to concentrations below those used to develop Table 6630:IV.

Method 6010A – *Table 1*

Recommended Wavelengths and Estimated Instrumental Detection Limits[a]

Element	Wavelength (nm)[a]	Estimated Detection Limit (µg/L)[b]
Aluminum	308.215	45
Antimony	206.833	32
Arsenic	193.696	53
Barium	455.403	2
Beryllium	313.042	0.3
Cadmium	226.502	4
Calcium	317.933	10
Chromium	267.716	7
Cobalt	228.616	7
Copper	324.754	6
Iron	259.940	7
Lead	220.353	42
Lithium	670.784	5
Mangesium	279.079	30
Manganese.	257.610	2
Molybdenum	202.030	8
Nickel	231.604	15
Phosphorus	213.618	51
Potassium	766.491	See note[c]
Selenium	196.026	75
Silicon	288.158	58
Silver	328.068	7
Sodium	588.995	29
Strontium	190.864	40
Thallium	190.864	40
Vanadium	292.402	8
Zinc	213.856	2

(a)   The wavelengths listed are recommended because of their sensitivity and overall acceptance. Other wavelengths may be substituted if they can provide the needed sensitivity and are treated with the same corrective techniques for spectral interference (see Paragraph 3.1). In time, other elements may be added as more information becomes available and as required.

(b)   The estimated instrumental detection limits shown are taken from Reference 1 in Section 10.0. They are given as a guide for an instrumental limit. The actual method detection limits are sample dependent and may vary as the sample matrix varies.

(c)   Highly dependent on operating conditions and plasma position.

Method 7000 Series – *Table 1*

Atomic Absorption Concentration Ranges

Metal	Direct Aspiration Detection Limit (mg/L)	Sensitivity (mg/L)	Furnace Procedure[a],[c] Detection Limit (µg/L)
Aluminum	0.1	1	
Antimony	0.2	0.5	3
Arsenic[b]	0.002		1
Barium	0.1	0.4	2
Beryllium	0.005	0.025	0.2
Cadmium	0.005	0.025	0.1
Calcium	0.01	0.08	
Chromium	0.05	0.25	1
Cobalt	0.05	0.2	1
Copper	0.02	0.1	1
Iron	0.03	0.12	1
Lead	0.1	0.5	1
Lithium	0.002	0.04	
Magnesium	0.001	0.007	
Manganese	0.01	0.05	0.2
Mercury[d]	0.0002		
Molybdenum(p)	0.1	0.4	1
Nickel	0.04	0.15	
Osmium	0.03	1	
Potassium	0.01	0.04	
Selenium[b]	0.002		2
Silver	0.01	0.06	0.2
Sodium	0.002	0.015	
Strontium	0.03	0.15	
Thallium	0.1	0.5	1
Tin	0.8	4	
Vanadium(p)	0.2	0.8	4
Zinc	0.005	0.02	0.05

NOTE: The symbol (p) indicates the use of pyrolytic graphite with the furnace procedure.

[a] For furnace sensitivity values, consult instrument operating manual.

[b] Gaseous hydride method

[c] The listed furnace values are those expected when using a 20-µL injection and normal gas flow, except in the cases of arsenic and selenium, where gas interrupt is used.

[d] Cold vapor technique

Method 8010A – *Table 1*

Chromatographic Conditions and Method Detection Limits for Halogenated Volatile Organics

Compound	CAS Registry Number	Retention Time (min)		Method Detection Limit[a] (µg/L)
		Col.1	Col.2	
Benzyl chloride[b]	100-44-7			
Bromobenzene	108-86-1	27.1		
Bromodichloromethane	75-27-4	13.7	14.6	0.1
Bromoform	75-25-2	19.2	19.2	0.2
Bromomethane	74-83-9	2.2	7.05	0.3
Carbon tetrachloride	56-23-5	13.0	14.4	0.12
Chlorobenzene	108-90-7	24.2	18.8	0.25
Chloroethane	75-00-3	3.33	8.68	0.52
2-Chloroethyl vinyl ether	110-75-8	18.0	[b]	0.13
Chloroform	67-66-3	10.7	12.1	0.05
Chloromethane	74-87-3	1.50	5.28	0.08
Dibromochloromethane	124-48-1	16.5	16.6	0.09
Dibromomethane	74-95-3	11.6	14.9	
1,2-Dichlorobenzene	95-50-1	34.9	23.5	0.15
1,3-Dichlorobenzene	541-73-1	34.0	22.4	0.32
1,4-Dichlorobenzene	106-46-7	35.5	22.3	0.24
Dichlorodifluoromethane[c]	75-71-8	2.62		
1,1-Dichloroethane	75-34-3	9.30	12.6	0.07
1,2-Dichloroethane	107-06-2	11.4	15.4	0.03
1,1-Dichloroethene	75-35-4	7.93	7.72	0.13
trans-1,2-Dichloroethene	156-60-5	10.1	9.38	0.10
Dichloromethane	75-09-2	5.25	10.1	[b]
1,2-Dichloropropane	78-87-5	14.9	16.6	0.04
cis-1,3-Dichloropropene	10061-01-5			
trans-1,3-Dichloropropene	10061-02-6	15.2	16.6	0.34
1,1,2,2-Tetrachloroethane	79-34-5	21.6		0.03
1,1,1,2-Tetrachloroethane	630-20-6	19.4	21.7	
Tetrachloroethene	127-18-4	21.7	15.0	0.03
1,1,1-Trichloroethane	71-55-6	12.6	13.1	0.03
1,1,2-Trichloroethane	79-00-5	16.5	18.1	0.02
Trichloroethene	79-01-6	15.8	13.1	0.12
Trichlorofluoromethane	75-69-4	7.18		
1,2,3-Trichloropropane	96-18-4	21.3		
Vinyl chloride	75-01-4	2.67	5.28	0.18

[a]   Using purge-and-trap method (Method 5030)
[b]   Demonstrated very erratic results when tested by purge and trap
[c]   See Section 4.10.2 of Method 5030 for guidance on selection of trapping material.

Determination of Estimated Quantitation Limits (EQL) for Various Matrices[a]

Matrix	Factor[b]
Ground water	10
Low-concentration soil	10
Water miscible liquid waste	500
High-concentration soil and sludge	1250
Non-water miscible waste	1250

[a]   Sample EQLs are highly matrix-dependent. The EQLs listed herein are provided for guidance and may not always be achievable.

[b]   EQL = [Method detection limit (Table 1)] X [Factor (Table 2)]. For non-aqueous samples, the factor is on a wet-weight basis.

Method 8010A – *Table 3*

Calibration and QC Acceptance Criteria[a]

Parameter	Range for Q (µg/L)	Limit for s (µg/L)	Range for x̄ (µg/L)	Range for P, P_s (%)
Bromodichloromethane	15.2-24.8	4.3	10.7-32.0	42-172
Bromoform	14.7-25.3	4.7	5.0-29.3	13-159
Bromomethane	11.7-28.3	7.6	3.4-24.5	D-144
Carbon tetrachloride	13.7-26.3	5.6	11.8-25.3	43-143
Chlorobenzene	14.4-25.6	5.0	10.2-27.4	38-150
Chloroethane	15.4-24.6	4.4	11.3-25.2	46-137
2-Chloroethylvinyl ether	12.0-28.0	8.3	4.5-35.5	14-186
Chloroform	15.0-25.0	4.5	12.4-24.0	49-133
Chloromethane	11.9-28.1	7.4	D-34.9	D-193
Dibromochloromethane	13.1-26.9	6.3	7.9-35.1	24-191
1,2-Dichlorobenzene	14.0-26.0	5.5	1.7-38.9	D-208
1,3-Dichlorobenzene	9.9-30.1	9.1	6.2-32.6	7-187
1,4-Dichlorobenzene	13.9-26.1	5.5	11.5-25.5	42-143
1,1-Dichloroethane	16.8-23.2	3.2	11.2-24.6	47-132
1,2-Dichloroethane	14.3-25.7	5.2	13.0-26.5	51-147
trans-1,2-Dichloroethane	12.8-27.2	6.4	11.4-27.1	38-155
1,1-Dichloroethene	12.6-27.4	6.6	10.2-27.3	28-167
Dichloromethane	15.5-24.5	4.0	7.0-27.6	25-162
1,2-Dichloropropane	14.8-25.2	5.2	10.1-29.9	44-156
cis-1,3-Dichloropropene	12.8-27.2	7.3	6.2-33.8	22-178
trans-1,3-Dichloropropene	12.8-27.2	7.3	6.2-33.8	22-178
1,1,2,2-Tetrachloroethane	9.8-30.2	9.2	6.6-31.8	8-184
Tetrachloroethene	14.0-26.0	5.4	8.1-29.6	26-162
1,1,1-Trichloroethane	14.2-25.8	4.9	10.8-24.8	41-138
1,1,2-Trichloroethane	15.7-24.3	3.9	9.6-25.4	39-136
Trichloroethene	15.4-24.6	4.2	9.2-26.6	35-146
Trichlorofluoromethane	13.3-26.7	6.0	7.4-28.1	21-156
Vinyl chloride	13.7-26.3	5.7	8.2-29.9	28-163

Q = Concentration measured in QC check sample, in µg/L
s = Standard deviation of four recovery measurements, in µg/L
x̄ = Average recovery for four recovery measurements, in µg/L
P, P_s = Percent recovery measured
D = Detected; result must be greater than zero.

[a] Criteria from 40 CFR Part 136 for Method 601 and were calculated assuming a QC check sample concentration of 20 µg/L.

Method 8020 – *Table 1*

Chromatographic Conditions and Method Detection Limits for Aromatic Volatile Organics

Compound	Retention Time (min)		Method Detection Limit[a] (μg/L)
	Col.1	Col.2	
Benzene	3.33	2.75	0.2
Chlorobenzene	9.17	8.02	0.2
1,4-Dichlorobenzene	16.8	16.2	0.3
1,3-Dichlorobenzene	18.2	15.0	0.4
1,2-Dichlorobenzene	25.9	19.4	0.4
Ethyl benzene	8.25	6.25	0.2
Toluene	5.75	4.25	0.2
Xylenes			

[a]   Using purge-and-trap method (Method 5030)

Method 8020 – *Table 3*

Calibration and QC Acceptance Criteria[a]

Parameter	Range for Q (μg/L)	Limit for s (mg/L)	Range for $\bar{x}$ (mg/L)	Range for P, P_s (%)
Benzene	15.4-24.6	4.1	10.0-27.9	39-150
Chlorobenzene	16.1-23.9	3.5	12.7-25.4	55-135
1,2-Dichlorobenzene	13.6-26.4	5.8	10.6-27.6	37-154
1,3-Dichlorobenzene	14.5-25.5	5.0	12.8-25.5	50-141
1,4-Dichlorobenzene	13.9-26.1	5.5	11.6-25.5	42-143
Ethylbenzene	12.6-27.4	6.7	10.0-28.2	32-160
Toluene	15.5-24.5	4.0	11.2-27.7	46-148

Q   =   Concentration measured in QC check sample, in μg/L

s   =   Standard deviation of four recovery measurements, in μg/L

$\bar{x}$ =   Average recovery for four recovery measurements, in μg/L

P, P_s =   Percent recovery measured

[a] Criteria from 40 CFR Part 136 for Method 602 and were calculated assuming a QC check sample concentration of 20 μg/L. These criteria are based directly upon the method performance data in Table 4. Where necessary, the limits for recovery have been broadened to assure applicability of the limits to concentrations below those used to develop Table 1.

Method 8021 – *Table 1*

Chromatographic Retention Times and Method Detection Limits (MDL) for Volatile Organic Compounds on Photoionization Detection (PID) and Hall Electrolytic Conductivity Detector (HECD) Detectors

Analyte	PID Ret. Time[a] minute	HECD Ret. Time minute	PID MDL µg/L	HECD MDL µg/L
Dichlorodifluoromethane	(b)	8.47		0.05
Chloromethane	(b)	9.47		0.03
Vinyl chloride	9.88	9.93	0.02	0.04
Bromomethane	(b)	11.95		1.1
Chloroethane	(b)	12.37		0.1
Trichlorofluoromethane	(b)	13.49		0.03
1,1-Dichloroethene	16.14	16.18	nd	0.07
Methylene chloride	(b)	18.39		0.02
trans-1,2-Dichloroethene	19.30	19.33	0.05	0.06
1,1-Dichloroethane	(b)	20.99		0.07
2,2-Dichloropropane	(b)	22.88		0.05
cis-1,2-Dichloroethene	23.11	23.14	0.02	0.01
Chloroform	(b)	23.64		0.02
Bromochloromethane	(b)	24.16		0.01
1,1,1-Trichloroethane	(b)	24.77		0.03
1,1-Dichloropropene	25.21	25.24	0.02	0.02
Carbon tetrachloride	(b)	25.47		0.01
Benzene	26.10	(b)	0.009	
1,2-Dichloroethane	(b)	26.27		0.03
Trichloroethene	27.99	28.02	0.02	0.01
1,2-Dichloropropane	(b)	28.66		0.006
Bromodichloromethane	(b)	29.43		0.02
Dibromomethane	(b)	29.59		2.2
Toluene	31.95	(b)	0.01	
1,1,2-Trichloroethane	(b)	33.21		nd
Tetrachloroethene	33.88	33.90	0.05	0.04
1,3-Dichloropropane	(b)	34.00		0.03
Dibromochloromethane	(b)	34.73		0.03
1,2-Dibromoethane	(b)	35.34		0.8
Chlorobenzene	36.56	36.59	0.003	0.01
Ethyl benzene	36.72	(b)	0.005	
1,1,1,2-Tetrachloroethane	(b)	36.80		0.005
m-Xylene	36.98	(b)	0.01	
p-Xylene	36.98	(b)	0.01	
o-Xylene	38.39	(b)	0.02	

Method 8021 – *Table 1*

Chromatographic Retention Times and Method Detection Limits (MDL) for Volatile Organic Compounds on Photoionization Detection (PID) and Hall Electrolytic Conductivity Detector (HECD) Detectors (Continued)

Analyte	PID Ret. Time[a] minute	HECD Ret. Time minute	PID MDL μg/L	HECD MDL μg/L
Styrene	38.57	(b)	0.01	
Isopropylbenzene	39.58	(b)	0.05	
Bromoform	(b)	39.75		1.6
1,1,2,2-Tetrachloroethane	(b)	40.35		0.01
1,2,3-Trichloropropane	(b)	40.81		0.4
n-Propylbenzene	40.87	(b)	0.004	
Bromobenzene	40.99	41.03	0.006	0.03
1,3,5-Trimethylbenzene	41.41	(b)	0.004	
2-Chlorotoluene	41.41	41.45	nd	0.01
4-Chlorotoluene	41.60	41.63	0.02	0.01
1,2,4-Trimethylbenzene	42.71	(b)	0.05	
tert-Butylbenzene	42.92	(b)	0.06	
sec-Butylbenzene	43.31	(b)	0.02	
4-Isopropyltoluene	43.81	(b)	0.01	
1,3-Dichlorobenzene	44.08	44.11	0.02	0.02
1,4-Dichlorobenzene	44.43	44.47	0.007	0.01
n-Butylbenzene	45.20	(b)	0.02	
1,2-Dichlorobenzene	45.71	45.74	0.05	0.02
1,2-Dibromo-3-chloropropane	(b)	48.57		3.0
1,2,4-Trichlorobenzene	51.43	51.46	0.02	0.03
Hexachlorobutadiene	51.92	51.96	0.06	0.02
Naphthalene	52.38	(b)	0.06	
1,2,3-Trichlorobenzene	53.34	53.37	nd	0.03

Internal Standards				
Analyte	PID Ret. Time[a] minute	HECD Ret. Time minute	PID MDL μg/L	HECD MDL μg/L
Fluorobenzene	26.84			
2-Bromo-1-Chloropropane		33.08		

[a] Retention times determined on 60 m x 0.75 mm ID VOCOL capillary column. Program: Hold 10 °C for 8 minutes, then program at 4 °C/min to 180 °C, and hold until all expected compounds eluted.

[b] Detector does not respond.

nd = Not determined.

Method 8021 – *Table 2*

Single Laboratory Accuracy and Precision Data for Volatile Organic Compounds in Water[c]

Analyte	Photoionizaton Detector		Hall Electrolytic Conductivity Detector	
	Recovery,[a] %	Standard Deviation of Recovery	Recovery,[a] %	Standard Deviation of Recovery
Benzene	99	1.2	(b)	(b)
Bromobenzene	99	1.7	97	2.7
Bromochloromethane	(b)	(b)	96	3.0
Bromodichloromethane	(b)	(b)	97	2.9
Bromoform	(b)	(b)	106	5.5
Bromomethane	(b)	(b)	97	3.7
n-Butylbenzene	100	4.4	(b)	(b)
sec-Butylbenzene	97	2.6	(b)	(b)
tert-Butylbenzene	98	2.3	(b)	(b)
Carbon tetrachloride	(b)	(b)	92	3.3
Chlorobenzene	100	1.0	103	3.7
Chloroethane	(b)	(b)	96	3.8
Chloroform	(b)	(b)	98	2.5
Chloromethane	(b)	(b)	96	8.9
2-Chlorotoluene	nd	nd	97	2.6
4-Chlorotoluene	101	1.0	97	3.1
1,2-Dibromo-3-chloropropane	(b)	(b)	86	9.9
Dibromochloromethane	(b)	(b)	102	3.3
1,2-Dibromoethane	(b)	(b)	97	2.7
Dibromomethane	(b)	(b)	109	7.4
1,2-Dichlorobenzene	102	2.1	100	1.5
1,3-Dichlorobenzene	104	1.7	106	4.3
1,4-Dichlorobenzene	103	2.2	98	2.3
Dichlorodifluoromethane	(b)	(b)	89	5.9
1,1-Dichloroethane	(b)	(b)	100	5.7
1,2-Dichloroethane	(b)	(b)	100	3.8
1,1-Dichloroethene	100	2.4	103	2.9
cis-1,2-Dichloroethene	nd	nd	105	3.5
trans-1,2-Dichloroethene	93	3.7	99	3.7
1,2-Dichloropropane	(b)	(b)	103	3.8
1,3-Dichloropropane	(b)	(b)	100	3.4
2,2-Dichloropropane	(b)	(b)	105	3.6
1,1-Dichloropropene	103	3.6	103	3.4
Ethyl benzene	101	1.4	(b)	(b)

Method 8021 – *Table 2*

Single Laboratory Accuracy and Precision Data for Volatile Organic Compounds in Water[c] (Continued)

Analyte	Photoionizaton Detector		Hall Electrolytic Conductivity Detector	
	Recovery,[a] %	Standard Deviation of Recovery	Recovery,[a] %	Standard Deviation of Recovery
Hexachlorobutadiene	99	9.5	98	8.3
Isopropylbenzene	98	0.9	(b)	(b)
4-Isopropyltoluene	98	2.4	(b)	(b)
Methylene chloride	(b)	(b)	97	2.8
Naphthalene	102	6.3	(b)	(b)
n-Propylbenzene	103	2.0	(b)	(b)
Styrene	104	1.4	(b)	(b)
1,1,1,2-Tetrachloroethane	(b)	(b)	99	2.3
1,1,2,2-Tetrachloroethane	(b)	(b)	99	6.8
Tetrachloroethene	101	1.8	97	2.4
Toluene	99	0.8	(b)	(b)
1,2,3-Trichlorobenzene	106	1.9	98	3.1
1,2,4-Trichlorobenzene	104	2.2	102	2.1
1,1,1-Trichloroethane	(b)	(b)	104	3.4
1,1,2-Trichloroethane	(b)	(b)	109	6.2
Trichloroethene	100	0.78	96	3.5
Trichlorofluoromethane	(b)	(b)	96	3.4
1,2,3-Trichloropropane	(b)	(b)	99	2.3
1,2,4-Trimethylbenzene	99	1.2	(b)	(b)
1,3,5-Trimethylbenzene	101	1.4	(b)	(b)
Vinyl chloride	109	5.4	95	5.6
o-Xylene	99	0.8	(b)	(b)
m-Xylene	100	1.4	(b)	(b)
p-Xylene	99	0.9	(b)	(b)

[a] Recoveries and standard deviations were determined from seven samples and spiked at 10 µg/L of each analyte. Recoveries were determined by internal standard method. Internal standards were: Fluorobenzene for PID; 2-Bromo-1-chloropropane for HECD.

[b] Detector does not respond.

[c] This method was tested in a single laboratory using water spiked at 10 µg/L (see reference 8).

nd = Not determined.

Method 8021 – *Table 3*

Determination of Estimated Quantitation Limits (EQL) for Various Matrices[a]

Matrix	Factor[b]
Ground water	10
Low-concentration soil	10
Water miscible liquid waste	500
High-concentration soil and sludge	1250
Non-water miscible waste	1250

[a] Sample EQLs are highly matrix-dependent. The EQLs listed herein are provided for guidance and may not always be achievable.

[b] EQL = [Method detection limit (Table 1)] X [Factor (Table 3)]. For non-aqueous samples, the factor is on a wet-weight basis.

Method 8080 – *Table 1*

Gas Chromatography of Pesticides and PCBs[a]

Compound	Retention Time (min) Col.1	Retention Time (min) Col.2	Method Detection Limit (μg/L)
Aldrin	2.40	4.10	0.004
α-BHC	1.35	1.82	0.003
β-BHC	1.90	1.97	0.006
δ-BHC	2.15	2.20	0.009
γ-BHC (Lindane)	1.70	2.13	0.004
Chlordane (technical)	e	e	0.014
4,4'-DDD	7.83	9.08	0.011
4,4'-DDE	5.13	7.15	0.004
4,4'-DDT	9.40	11.75	0.012
Dieldrin	5.45	7.23	0.002
Endosulfan I	4.50	6.20	0.014
Endosulfan II	8.00	8.28	0.004
Endosulfan sulfate	14.22	10.70	0.066
Endrin	6.55	8.10	0.006
Endrin aldehyde	11.82	9.30	0.023
Heptachlor	2.00	3.35	0.003
Heptachlor epoxide	3.50	5.00	0.083
Methoxychlor	18.20	26.60	0.176
Toxaphene	e	e	0.24
PCB-1016	e	e	nd
PCB-1221	e	e	nd
PCB-1232	e	e	nd
PCB-1242	e	e	0.065
PCB-1248	e	e	nd
PCB-1254	e	e	nd
PCB-1260	e	e	nd

[a] U.S. EPA. Method 617. Organochloride Pesticides and PCBs. Environmental Monitoring and Support Laboratory, Cincinnati, Ohio 45268

e = Multiple peak response
nd = not determined

Method 8080 – *Table 2*

Determination of Practical Quantitation Limits (PQL) for Various Matrices[a]

Matrix	Factor[b]
Ground water	10
Low-concentration soil by sonication with GPC cleanup	670
High-concentration soil and sludges by sonication	10000
Non-water miscible waste	100000

[a] Sample PQLs are highly matrix-dependent. The PQLs listed herein are provided for guidance and may not always be achievable.

[b] PQL = [Method detection limit (Table 1)] X [Factor (Table 2)]. For non-aqueous samples, the factor is on a wet-weight basis.

Method 8080 – *Table 3*

QC Acceptance Criteria[a]

Parameter	Test conc. (μg/L)	Limit for s (μg/L)	Range for x̄ (μg/L)	Range for P, P_s (%)
Aldrin	2.0	0.42	1.08-2.24	42-122
α-BHC	2.0	0.48	0.98-2.44	37-134
β-BHC	2.0	0.64	0.78-2.60	17-147
δ-BHC	2.0	0.72	1.01-2.37	19-140
γ-BHC	2.0	0.46	0.86-2.32	32-127
Chlordane	50	10.0	27.6-54.3	45-119
4,4'-DDD	10	2.8	4.8-12.6	31-141
4,4'-DDE	2.0	0.55	1.08-2.60	30-145
4,4'-DDT	10	3.6	4.6-13.7	25-160
Dieldrin	2.0	0.76	1.15-2.49	36-146
Endosulfan I	2.0	0.49	1.14-2.82	45-153
Endosulfan II	10	6.1	2.2-17.1	D-202
Endosulfan sulfate	10	2.7	3.8-13.2	26-144
Endrin	10	3.7	5.1-12.6	30-147
Heptachlor	2.0	0.40	0.86-2.00	34-111
Heptachlor epoxide	2.0	0.41	1.13-2.63	37-142
Toxaphene	50	12.7	27.8-55.6	41-126
PCB-1016	50	10.0	30.5-51.5	50-114
PCB-1221	50	24.4	22.1-75.2	15-178
PCB-1232	50	17.9	14.0-98.5	10-215
PCB-1242	50	12.2	24.8-69.6	39-150
PCB-1248	50	15.9	29.0-70.2	38-158
PCB-1254	50	13.8	22.2-57.9	29-131
PCB-1260	50	10.4	18.7-54.9	8-127

s   =   Standard deviation of four recovery measurements, in μg/L

x̄   =   Average recovery for four recovery measurements, in μg/L

P, P_s =   Percent recovery measured

D   =   Detected; result must be greater than zero.

[a] Criteria from 40 CFR Part 136 for Method 608. These criteria are based directly upon the method performance data in Table 4. Where necessary, the limits for recovery have been broadened to assure applicability of the limits to concentrations below those used to develop Table 4.

Method 8150A – *Table 1*

Chromatographic Conditions and Detection Limits for Chlorinated Herbicides

Compound	Retention time (min)[a]				Method Detection
	Col.1a	Col.1b	Col. 2	Col. 3	Limit (µg/L)
2,4-D	2.0		1.6		1.2
2,4-DB	4.1				0.91
2,4,5-T	3.4		2.4		0.20
2,4,5-TP (Silvex)	2.7		2.0		0.17
Dalapon				5.0	5.8
Dicamba	1.2		1.0		0.27
Dichloroprop		4.8			0.65
Dinoseb		11.2			0.07
MCPA		4.1			249
MCPP		3.4			192

[a] Column conditions are given in Sections 4.1 and 7.5

Method 8150A – *Table 2*

Determination of Estimated Quantitation Limits (EQL) for Various Matrices[a]

Matrix	Factor[b]
Groundwater (based on one liter sample size)	10
Soil/sediment and other solids	200
Waste samples	100000

[a] Sample EQLs are highly matrix-dependent. The EQLs listed herein are provided for guidance and may not always be achievable.

[b] EQL = [Method detection limit (Table 1)] X [Factor (Table 2)]. For non-aqueous samples, the factor is on a wet-weight basis.

Method 8150A – *Table 3*

Single-Operator Accuracy and Precision[a]

Parameter	Sample Type	Spike (µg/L)	Mean Recovery (%)	Standard Deviation (%)
2,4-D	DW	10.9	75	4
	MW	10.1	77	4
	MW	200	65	5
Dalapon	DW	23.4	66	8
	MW	23.4	96	13
	MW	468	81	9
2,4-DB	DW	10.3	93	3
	MW	10.4	93	3
	MW	208	77	6
Dicamba	DW	1.2	79	7
	MW	1.1	86	9
	MW	22.2	82	6
Dichlorprop	DW	10.7	97	2
	MW	10.7	72	3
	MW	213	100	2
Dinoseb	MW	0.5	86	4
	MW	102	81	3
MCPA	DW	2020	98	4
	MW	2020	73	3
	MW	21400	97	2
MCPP	DW	2080	94	4
	MW	2100	97	3
	MW	20440	95	2
2,4,5-T	DW	1.1	85	6
	MW	1.3	83	4
	MW	25.5	78	5
2,4,5-TP	DW	1.0	88	5
	MW	1.3	88	4
	MW	25.0	72	5

[a] All results based upon seven replicate analyses. Esterification performed using the bubbler method. Data obtained from reference 9.

DW = ASTM Type II

MW = Municipal water

Method 8240A – *Table 2*

Estimated Quantitation Limits (EQL) for Volatile Organics[a]

Volatiles	CAS Number	Estimated Quantitation Limits[b]	
		Ground Water µg/L	Low Soil/ Sediment µg/kg
Acetone	67-64-1	100	100
Acetonitrile	75-08-8	100	100
Allyl chloride	107-05-1	5	5
Benzene	71-43-2	5	5
Benzyl chloride	100-44-7	100	100
Bromodichloromethane	75-27-4	5	5
Bromoform	75-25-2	5	5
Bromomethane	74-83-9	10	10
2-Butanone	78-93-3	100	100
Carbon disulfide	75-15-0	100	100
Carbon tetrachloride	56-23-5	5	5
Chlorobenzene	108-90-7	5	5
Chlorodibromomethane	124-48-1	5	5
Chloroethane	75-00-3	10	10
2-Chloroethyl vinyl ether	110-75-8	10	10
Chloroform	67-66-3	5	5
Chloromethane	74-87-3	10	10
Chloroprene	126-99-8	5	5
1,2-Dibromo-3-chloropropane	96-12-8	100	100
1,2-Dibromoethane	106-93-4	5	5
Dibromomethane	74-95-3	5	5
1,4-Dichloro-2-butene	764-41-0	100	100
Dichlorodiflouromethane	75-71-8	5	5
1,1-Dichloroethane	75-34-3	5	5
1,2-Dichloroethane	107-06-2	5	5
1,1-Dichloroethene	75-35-4	5	5
trans-1,2-Dichloroethene	156-60-5	5	5
1,2-Dichloropropane	78-87-5	5	5
cis-1,3-Dichloropropene	10061-01-5	5	5
trans-1,3-Dichloropropene	10061-02-6	5	5
Ethylbenzene	100-41-4	5	5
Ethyl methacrylate	97-63-2	5	5
2-Hexanone	591-78-6	50	50

Method 8240A – *Table 2*

Estimated Quantitation Limits (EQL) for Volatile Organics[a] (Continued)

Volatiles	CAS Number	Estimated Quantitation Limits[b]	
		Ground Water µg/L	Low Soil/ Sediment µg/kg
Isobutyl alcohol	78-83-1	100	100
Methacrylonitrile	109-77-3	100	100
Methylene chloride	75-09-2	5	5
Methyl iodide	74-88-4	5	5
Methyl methacrylate	80-62-6	5	50
4-Methyl-2-pentanone	108-10-1	50	50
Pentachloroethane	76-01-7	10	10
Proprionitrile	107-12-0	100	100
Styrene	100-42-5	5	5
1,1,1,2-Tetrachloroethane	630-20-6	5	5
1,1,2,2-Tetrachloroethane	79-34-5	5	5
Tetrachloroethene	127-18-4	5	5
Toluene	108-88-3	5	5
1,1,1-Trichloroethane	71-55-6	5	5
1,1,2-Trichloroethane	79-00-5	5	5
Trichloroethene	79-01-6	5	5
1,2,3-Trichloropropane	107-12-0	5	5
Vinyl acetate	108-05-4	50	50
Vinyl chloride	75-01-4	10	10
Xylenes (Total)	1330-20-7	5	5

Factor[c]	Other Matrices
50	Water miscible liquid waste
125	High- concentration soil and sludge
500	Non-water miscible waste

[a] Sample EQLs are highly matrix-dependent. The EQLs listed herein are provided for guidance and may not always be achievable. See the following information for further guidance on matrix-dependant EQLs.

[b] EQLs listed for soil/sediment are based on wet weight. Normally data is reported on a dry weight basis; therefore, EQLs will be higher based on the % dry weight in each sample.

[c] EQL = [EQL for low soil sediment (Table 2) X [Factor]. For non-aqueous samples, the factor is on a wet weight basis.

Method 8240A – *Table 3*

BFB KEY Ion Abundance Criteria

Mass	Ion Abundance Criteria
50	15 to 40% of mass 95
75	30 to 60% of mass 95
95	base peak, 100% relative abundance
96	5 to 9% of mass 95
173	less than 2% of mass 174
174	greater than 50% of mass 95
175	5 to 9% of mass 174
176	greater than 95% but less than 101% of mass 174
177	5 to 9% of mass 176

Method 8240A – *Table 6*

Calibration and QC Acceptance Criteria[a]

Parameter	Range for Q (µg/L)	Limit for s (µg/L)	Range for x̄ (µg/L)	Range for P, Ps (%)
Benzene	12.8-27.2	6.9	15.2-26.0	37-151
Bromodichloromethane	13.1-26.9	6.4	10.1-28.0	35-155
Bromoform	14.2-25.8	5.4	11.4-31.1	45-169
Bromomethane	2.8-37.2	17.9	D-41.2	D-242
Carbon tetrachloride	14.6-25.4	5.2	17.2-23.5	70-140
Chlorobenzene	13.2-26.8	6.3	16.4-27.4	37-160
2-Chloroethylvinyl ether	D-44.8	25.9	D-50.4	D-305
Chloroform	13.5-26.5	6.1	13.7-24.2	51-138
Chloromethane	D-40.8	19.8	D-45.9	D-273
Dibromochloromethane	13.5-26.5	6.1	13.8-26.6	53-149
1,2-Dichlorobenzene	12.6-27.4	7.1	11.8-34.7	18-190
1,3-Dichlorobenzene	14.6-25.4	5.5	17.0-28.8	59-156
1,4-Dichlorobenzene	12.6-27.4	7.1	11.8-34.7	18-190
1,1-Dichloroethane	14.5-25.5	5.1	14.2-28.4	59-155
1,2-Dichloroethane	13.6-26.4	6.0	14.3-27.4	49-155
1,1-Dichloroethene	10.1-29.9	9.1	3.7-42.3	D-234
trans-1,2-Dichloroethene	13.9-26.1	5.7	13.6-28.4	54-156
1,2-Dichloropropane	6.8-33.2	13.8	3.8-36.2	D-210
cis-1,3-Dichloropropene	4.8-35.2	15.8	1.0-39.0	D-227
trans-1,3-Dichloropropene	10.0-30.0	10.4	7.6-32.4	17-183
Ethyl benzene	11.8-28.2	7.5	17.4-26.7	37-162
Methylene chloride	12.1-27.9	7.4	D-41.0	D-221
1,1,2,2-Tetrachloroethane	12.1-27.9	7.4	13.5-27.2	46-157
Tetrachloroethene	14.7-25.3	5.0	17.0-26.6	64-148
Toluene	14.9-25.1	4.8	16.6-26.7	47-150
1,1,1-Trichloroethane	15.0-25.0	4.6	13.7-30.1	52-162
1,1,2-Trichloroethane	14.2-25.8	5.5	14.3-27.1	52-150
Trichloroethene	13.3-26.7	6.6	18.5-27.6	71-157
Trichlorofluoromethane	9.6-30.4	10.0	8.9-31.5	17-181
Vinyl chloride	0.8-39.2	20.0	D-43.5	D-251

Q = Concentration measured in QC check sample, in µg/L
s = Standard deviation of four recovery measurements, in µg/L
x̄ = Average recovery for four recovery measurements, in µg/L
P, Ps = Percent recovery measured
D = Detected; result must be greater than zero.

[a] Criteria from 40 CFR Part 136 for Method 624 and were calculated assuming a QC check sample concentration of 20 µg/L. These criteria are based directly upon the method performance data in Table 7, Method 8240A, EPA SW-846. Where necessary, the limits for recovery have been broadened to assure applicability of the limits to concentrations below those used to develop Table 7.

Method 8240A – *Table 8*

Surrogate Spike Recovery Limits For Water and Soil/Sediment Samples

Surrogate Compound	Low/High Water	Low/High Soil/Sediment
4-Bromofluorobenzene	86-115	74-121
1,2-Dichloroethane-d$_4$	76-114	70-121
Toluene-d$_8$	88-110	81-117

Method 8260 – *Table 1*

Chromatographic Retention Times and Method Detection Limits (MDL) for Volatile Organic Compounds on Wide Bore Capillary Columns.

Analyte	Retention Time (minutes)			MDL[d] (µg/L)
	Column 1[a]	Column 2[b]	Column 2'[c]	
Dichlorodifluoromethane	1.55	0.70	3.13	0.10
Chloromethane	1.63	0.73	3.13	0.10
Vinyl chloride	1.71	0.79	3.93	0.17
Bromomethane	2.01	0.96	4.80	0.11
Chloroethane	2.09	1.02	--	0.10
Trichlorofluoromethane	2.27	1.19	6.20	0.08
1,1-Dichloroethene	2.89	1.57	7.83	0.12
Methylene chloride	3.60	2.06	9.27	0.03
trans-1,2-Dichloroethene	3.98	2.36	9.90	0.06
1,1-Dichloroethane	4.85	2.93	10.80	0.04
2,2-Dichloropropane	6.01	3.80	11.87	0.35
cis-1,2-Dichloroethene	6.19	3.90	11.93	0.12
Chloroform	6.40	4.80	12.60	0.03
Bromochloromethane	6.74	4.38	12.37	0.04
1,1,1-Trichloroethane	7.27	4.84	12.83	0.08
Carbon tetrachloride	7.61	5.26	13.17	0.21
1,1-Dichloropropene	7.68	5.29	13.10	0.10
Benzene	8.23	5.67	13.50	0.04
1,2-Dichloroethane	8.40	5.83	13.63	0.06
Trichloroethene	9.59	7.27	14.80	0.19
1,2-Dichloropropane	10.09	7.66	15.20	0.04
Bromodichloromethane	10.59	8.49	15.80	0.08
Dibromomethane	10.65	7.93	15.43	0.24
cis-1,3-Dichloropropene	--	--	17.90	--
Toluene	12.43	10.00	17.40	0.11
trans-1,3-Dichloropropene	--	--	16.70	--
1,1,2-Trichloroethane	13.41	11.05	18.30	0.10
Tetrachloroethene	13.74	11.15	18.60	0.14
1,3-Dichloropropane	14.04	11.31	18.70	0.04
Dibromochloromethane	14.39	11.85	19.20	0.05
1,2-Dibromoethane	14.73	11.83	19.40	0.06
1-Chlorohexane	15.46	13.29	--	0.05
Chlorobenzene	15.76	13.01	20.67	0.04
1,1,1,2-Tetrachloroethane	15.94	13.33	20.87	0.05
Ethyl benzene	15.99	13.39	21.00	0.06
p-Xylene	16.12	13.69	21.30	0.13
m-Xylene	16.17	13.68	21.37	0.05
o-Xylene	17.11	14.52	22.27	0.11
Styrene	17.31	14.60	22.40	0.04
Bromoform	17.93	14.88	22.77	0.12
Isopropylbenzene	18.06	15.46	23.30	0.15

Method 8260 – *Table 1*

Chromatographic Retention Times and Method Detection Limits (MDL) for Volatile Organic Compounds on Wide Bore Capillary Columns. (Continued)

| Analyte | Retention Time (minutes) | | | MDL[d] |
	Column 1[a]	Column 2[b]	Column 2'[c]	(µg/L)
1,1,2,2-Tetrachloroethane	18.72	16.35	24.07	0.04
Bromobenzene	18.95	15.86	24.00	0.03
1,2,3-Trichloropropane	19.02	15.86	24.00	0.32
n-Propylbenzene	19.06	16.41	24.33	0.04
2-Chlorotoluene	19.34	16.42	24.53	0.04
1,3,5-Trimethylbenzene	19.47	16.90	24.83	0.05
4-Chlorotoluene	19.50	16.72	24.77	0.06
tert-Butylbenzene	20.28	17.57	26.60	0.14
1,2,4-Trimethylbenzene	20.34	17.70	31.50	0.13
sec-Butylbenzene	20.79	18.09	26.13	0.13
p-Isopropyltoluene (4-Isopropyl toluene)	21.20	18.52	26.50	0.12
1,3-Dichlorobenzene	21.22	18.14	26.37	0.12
1,4-Dichlorobenzene	21.55	18.39	26.60	0.03
n-Butylbenzene	22.22	19.49	27.32	0.11
1,2-Dichlorobenzene	22.52	19.17	27.43	0.03
1,2-Dibromo-3-chloropropane	24.53	21.08	--	0.26
1,2,4-Trichlorobenzene	26.55	23.08	31.50	0.04
Hexachlorobutadiene	26.99	23.68	32.07	0.11
Naphthalene	27.17	23.52	32.20	0.04
1,2,3-Trichlorobenzene	27.78	24.18	32.97	0.03

| Internal Standards/Surrogates | | | | |
| Analyte | Retention Time (minutes) | | | MDL[d] |
	Column 1[a]	Column 2[b]	Column 2'[c]	(µg/L)
4-Bromofluorobenzene	18.63	15.71	23.63	

[a] Column 1 - 60 meter x 0.75 mm ID VOCOL Capillary. Hold at 10 °C for 5 minutes, then program to 160 °C at 6 °C/min.

[b] Column 2 - 30 meter x 0.53 mm ID DB-624 wide-bore capillary using cryogenic oven. Hold at 10 °C for 5 minutes, then program to 160 °C at 6 °C/min.

[c] Column 2' - 30 meter x 0.53 ID DB-624 wide-bore capillary, cooling GC oven to ambient temperatures. Hold at 10 °C for 6 minutes, program to 70 °C at 10 °C/min, program to 120 °C at 5 °C/min, then program to 180 °C at 8 °C/min.

[d] MDL based on a 25 mL sample volume.

Method 8260 – *Table 2*

Chromatographic Retention Times and Method Detection Limits (MDL) for Volatile Organic Compounds on Narrow Bore Capillary Columns.

Analyte	Retention Time (min.) Column 3[a]	MDL[b] (µg/L)
Dichlorodifluoromethane	0.88	0.11
Chloromethane	0.97	0.05
Vinyl chloride	1.04	0.04
Bromomethane	1.29	0.06
Chloroethane	1.45	0.02
Trichlorofluoromethane	1.77	0.07
1,1-Dichloroethene	2.33	0.05
Methylene chloride	2.66	0.09
trans-1,2-Dichloroethene	3.54	0.03
1,1-Dichloroethane	4.03	0.03
cis-1,2-Dichloroethene	5.07	0.06
2,2-Dichloropropane	5.31	0.08
Chloroform	5.55	0.04
Bromochloromethane	5.63	0.09
1,1,1-Trichloroethane	6.76	0.04
1,2-Dichloroethane	7.00	0.02
1,1-Dichloropropene	7.16	0.12
Carbon tetrachloride	7.41	0.02
Benzene	7.41	0.03
1,2-Dichloropropane	8.94	0.02
Trichloroethene	9.02	0.02
Dibromomethane	9.09	0.01
Bromodichloromethane	9.34	0.03
Toluene	11.51	0.08
1,1,2-Trichloroethane	11.99	0.08
1,3-Dichloropropane	12.48	0.08
Dibromochloromethane	12.80	0.07
Tetrachloroethene	13.20	0.05
1,2-Dibromoethane	13.60	0.10
Chlorobenzene	14.33	0.03
1,1,1,2-Tetrachloroethane	14.73	0.07
Ethyl benzene	14.73	0.03
p-Xylene	15.30	0.06
m-Xylene	15.30	0.03
Bromoform	15.70	0.20
o-Xylene	15.78	0.06
Styrene	15.78	0.27
1,1,2,2-Tetrachloroethane	15.78	0.20
1,2,3-Trichloropropane	16.26	0.09
Isopropylbenzene	16.42	0.10
Bromobenzene	16.42	0.11

Chromatographic Retention Times and Method Detection Limits (MDL) for Volatile Organic Compounds on Narrow Bore Capillary Columns. (Continued)

Analyte	Retention Time (min.)	
	Column 3[a]	MDL[b] (µg/L)
2-Chlorotoluene	16.74	0.08
n-Propylbenzene	16.82	0.10
4-Chlorotoluene	16.82	0.06
1,3,5-Trimethylbenzene	16.99	0.06
tert-Butylbenzene	17.31	0.33
1,2,4-Trimethylbenzene	17.31	0.09
sec-Butylbenzene	17.47	0.12
1,3-Dichlorobenzene	17.47	0.05
p-Isopropyltoluene (4-Isopropyl toluene)	17.63	0.26
1,4-Dichlorobenzene	17.63	0.04
1,2-Dichlorobenzene	17.79	0.05
n-Butylbenzene	17.95	0.10
1,2-Dibromo-3-chloropropane	18.03	0.50
1,2,4-Trichlorobenzene	18.84	0.20
Naphthalene	19.09	0.10
Hexachlorobutadiene	19.24	0.10
1,2,3-Trichlorobenzene	19.24	0.14

[a] Column 3 - 30 meter x 0.32 mm ID DB5 Capillary with 1 µm film thickness.

[b] MDL based on a 25 mL sample volume.

Method 8260 – *Table 3*

ESTIMATED QUANTITATION LIMITS FOR VOLATILE ANALYTES[a]

	Estimated Quatitation Limits		
	Ground water µg/L		Low Soil/Sediment[b] µg/kg
Volume of water purged	5 mL	25 mL	
All analytes in Table 1	5	1	5

Other Matrices	Factor[c]
Water miscible liquid waste	50
High-concentration soil and sludge	125
Non-water miscible waste	500

[a] Estimated Quantitation Limit (EQL) - The lowest concentration that can be reliably achieved within specified limits of precision and accuracy during routine laboratory operating conditions. The EQL is generally 5 to 10 times the MDL. However, it may be nominally chosen within these guidelines to simplify data reporting. For many analytes the EQL analyte concentration is selected for the lowest non-zero standard in the calibration curve. Sample EQLs are highly matrix-dependent. The EQLs listed herein are provided for guidance and may not always be achievable. See the following information for further guidance on matrix-dependent EQLs.

[b] EQLs listed for soil/sediment are based on wet weight. Normally data is reported on a dry weight basis; therefore, EQLs will be higher, based on the percent dry weight in each sample.

[c] EQL = [EQL for low soil sediment (table 3)] X [Factor}. For non-aqueous samples, the factor is on a wet-weight basis.

Method 8260 – *Table 4*

BFB Mass - Intensity Specifications (4-Bromofluorobenzene)

Mass	Intensity Required (relative abundance)
50	15 to 40% of mass 95
75	30 to 60% of mass 95
95	base peak, 100% relative abundance
96	5 to 9% of mass 95
173	less than 2% of mass 174
174	greater than 50% of mass 95
175	5 to 9% of mass 174
176	greater than 95% but less than 101% of mass 174
177	5 to 9% of mass 176

Method 8260 – *Table 7*

Single Laboratory Accuracy and Precision Data for Volatile Organic Compounds in Water Determined with a Wide Bore Capillary Column

Analyte	Conc. Range, µg/L	Number of Samples	Recovery,[a] %	Standard Deviation of Recovery[b]	Percent Rel. Std. Dev.
Benzene	0.1 - 10	31	97	6.5	5.7
Bromobenzene	0.1 - 10	30	100	5.5	5.5
Bromochloromethane	0.5 - 10	24	90	5.7	6.4
Bromodichloromethane	0.1 - 10	30	95	5.7	6.1
Bromoform	0.5 - 10	18	101	6.4	6.3
Bromomethane	0.5 - 10	18	95	7.8	8.2
n-Butylbenzene	0.5 - 10	18	100	7.6	7.6
sec-Butylbenzene	0.5 - 10	16	100	7.6	7.6
tert-Butylbenzene	0.5 - 10	18	102	7.4	7.3
Carbon tetrachloride	0.5 - 10	24	84	7.4	8.8
Chlorobenzene	0.1 - 10	31	98	5.8	5.9
Chloroethane	0.5 - 10	24	89	8.0	9.0
Chloroform	0.5 - 10	24	90	5.5	6.1
Chloromethane	0.5 - 10	23	93	8.3	8.9
2-Chlorotoluene	0.1 - 10	31	90	5.6	6.2
4-Chlorotoluene	0.1 - 10	31	99	8.2	8.3
1,2-Dibromo-3-chloropropane	0.5 - 10	24	83	16.6	19.9
Dibromochloromethane	0.1 - 10	31	92	6.5	7.0
1,2-Dibromoethane	0.5 - 10	24	102	4.0	3.9
Dibromomethane	0.5 - 10	24	100	5.6	5.6
1,2-Dichlorobenzene	0.1 - 10	31	93	5.8	6.2
1,3-Dichlorobenzene	0.5 - 10	24	99	6.8	6.9
1,4-Dichlorobenzene	0.2 - 20	31	103	6.6	6.4
Dichlorodifluoromethane	0.5 - 10	18	90	6.9	7.7
1,1-Dichloroethane	0.5 - 10	24	96	5.1	5.3
1,2-Dichloroethane	0.1 - 10	31	95	5.1	5.4
1,1-Dichloroethene	0.1 - 10	34	94	6.3	6.7
cis-1,2-Dichloroethene	0.5 - 10	18	101	6.7	6.7
trans-1,2-Dichloroethene	0.1 - 10	30	93	5.2	5.6
1,2-Dichloropropane	0.1 - 10	30	97	5.9	6.1
1,3-Dichloropropane	0.1 - 10	31	96	5.7	6.0
2,2-Dichloropropane	0.5 - 10	12	86	14.6	16.9
1,1-Dichloropropene	0.5 - 10	18	98	8.7	8.9
Ethyl benzene	0.1 - 10	31	99	8.4	8.6
Hexachlorobutadiene	0.5 - 10	18	100	6.8	6.8
Isopropylbenzene	0.5 - 10	16	101	7.7	7.6
p-Isopropyltoluene (4-Isopropyl toluene)	0.1 - 10	23	99	6.7	6.7
Methylene chloride	0.1 - 10	30	95	5.0	5.3
Naphthalene	0.1 - 100	31	104	8.6	8.2
n-Propylbenzene	0.1 - 10	31	100	5.8	5.8

Method 8260 – *Table 7*

Single Laboratory Accuracy and Precision Data for Volatile Organic Compounds in Water Determined with a Wide Bore Capillary Column (Continued)

Analyte	Conc. Range, µg/L	Number of Samples	Recovery,(a) %	Standard Deviation of Recovery(b)	Percent Rel. Std. Dev.
Styrene	0.1 - 100	39	102	7.3	7.2
1,1,1,2-Tetrachloroethane	0.5 - 10	24	90	6.1	6.8
1,1,2,2-Tetrachloroethane	0.1 - 10	30	91	5.7	6.3
Tetrachloroethene	0.5 - 10	24	89	6.0	6.8
Toluene	0.5 - 10	18	102	8.1	8.0
1,2,3-Trichlorobenzene	0.5 - 10	18	109	9.4	8.6
1,2,4-Trichlorobenzene	0.5 - 10	18	108	9.0	8.3
1,1,1-Trichloroethane	0.5 - 10	18	98	7.9	8.1
1,1,2-Trichloroethane	0.5 - 10	18	104	7.6	7.3
Trichloroethene	0.5 - 10	24	90	6.5	7.3
Trichlorofluoromethane	0.5 - 10	24	89	7.2	8.1
1,2,3-Trichloropropane	0.5 - 10	16	108	15.6	14.4
1,2,4-Trimethylbenzene	0.5 - 10	18	99	8.0	8.1
1,3,5-Trimethylbenzene	0.5 - 10	23	92	6.8	7.4
Vinyl chloride	0.5 - 10	18	98	6.5	6.7
o-Xylene	0.1 - 31	18	103	7.4	7.2
m-Xylene	0.1 - 10	31	97	6.3	6.5
p-Xylene	0.5 - 10	18	104	8.0	7.7

(a)   Recoveries were calculated using internal standard method. Internal standard was fluorobenzene.

(b)   Standard deviation was calculated by pooling data from three concentrations.

Method 8260 – *Table 8*

Single Laboratory Accuracy and Precision Data for Volatile Organic Compounds in Water Determined with a Narrow Bore Capillary Column

Analyte	Conc. Range, µg/L	Number of Samples	Recovery,[a] %	Standard Deviation of Recovery	Percent Rel. Std. Dev.
Benzene	0.1	7	99	6.2	6.3
Bromobenzene	0.5	7	97	7.4	7.6
Bromochloromethane	0.5	7	97	5.8	6.0
Bromodichloromethane	0.1	7	100	4.6	4.6
Bromoform	0.5	7	101	5.4	5.3
Bromomethane	0.5	7	99	7.1	7.2
n-Butylbenzene	0.5	7	94	6.0	6.4
sec-Butylbenzene	0.5	7	110	7.1	6.5
tert-Butylbenzene	0.5	7	110	2.5	2.3
Carbon tetrachloride	0.1	7	108	6.8	6.3
Chlorobenzene	0.1	7	91	5.8	6.4
Chloroethane	0.1	7	100	5.8	5.8
Chloroform	0.1	7	105	3.2	3.0
Chloromethane	0.5	7	101	4.7	4.7
2-Chlorotoluene	0.5	7	99	4.6	4.6
4-Chlorotoluene	0.5	7	96	7.0	7.3
1,2-Dibromo-3-chloropropane	0.5	7	92	10.0	10.9
Dibromochloromethane	0.1	7	99	5.6	5.7
1,2-Dibromoethane	0.5	7	97	5.6	5.8
Dibromomethane	0.5	7	93	5.6	6.0
1,2-Dichlorobenzene	0.1	7	97	3.5	3.6
1,3-Dichlorobenzene	0.1	7	101	6.0	5.9
1,4-Dichlorobenzene	0.1	7	106	6.5	6.1
Dichlorodifluoromethane	0.1	7	99	8.8	8.9
1,1-Dichloroethane	0.5	7	98	6.2	6.3
1,2-Dichloroethane	0.1	7	100	6.3	6.3
1,1-Dichloroethene	0.1	7	95	9.0	9.5
cis-1,2-Dichloroethene	0.1	7	100	3.7	3.7
trans-1,2-Dichloroethene	0.1	7	98	7.2	7.3
1,2-Dichloropropane	0.5	7	96	6.0	6.3
1,3-Dichloropropane	0.5	7	99	5.8	5.9
2,2-Dichloropropane	0.5	7	99	4.9	4.9
1,1-Dichloropropene	0.5	7	102	7.4	7.3
Ethyl benzene	0.5	7	99	5.2	5.3
Hexachlorobutadiene	0.5	7	100	6.7	6.7
Isopropylbenzene	0.5	7	102	6.4	6.3
p-Isopropyltoluene (4-Isopropyl toluene)	0.5	7	113	13.0	11.5
Methylene chloride	0.5	7	97	13.0	13.4
Naphthalene	0.5	7	98	7.2	7.3
n-Propylbenzene	0.5	7	99	6.6	6.7

Method 8260 – *Table 8*

Single Laboratory Accuracy and Precision Data for Volatile Organic Compounds in Water Determined with a Narrow Bore Capillary Column (Continued)

Analyte	Conc. Range, μg/L	Number of Samples	Recovery,[a] %	Standard Deviation of Recovery	Percent Rel. Std. Dev.
Styrene	0.5	7	96	19.0	19.8
1,1,1,2-Tetrachloroethane	0.5	7	100	4.7	4.7
1,1,2,2-Tetrachloroethane	0.5	7	100	12.0	12.0
Tetrachloroethene	0.1	7	96	5.0	5.2
Toluene	0.5	7	100	5.9	5.9
1,2,3-Trichlorobenzene	0.5	7	102	8.9	8.7
1,2,4-Trichlorobenzene	0.5	7	91	16.0	17.6
1,1,1-Trichloroethane	0.5	7	100	4.0	4.0
1,1,2-Trichloroethane	0.5	7	102	4.9	4.8
Trichloroethene	0.1	7	104	2.0	1.9
Trichlorofluoromethane	0.1	7	97	4.6	4.7
1,2,3-Trichloropropane	0.5	7	96	6.5	6.8
1,2,4-Trimethylbenzene	0.5	7	96	6.5	6.8
1,3,5-Trimethylbenzene	0.5	7	101	4.2	4.2
Vinyl chloride	0.1	7	104	0.2	0.2
o-Xylene	0.5	7	106	7.5	7.1
m-Xylene	0.5	7	106	4.6	4.3
p-Xylene	0.5	7	97	6.1	6.3

[a]   Recoveries were calculated using internal standard method. Internal standard was fluorobenzene.

Surrogate Spike Recovery Limits for Water and Soil/Sediment Samples

Surrogate Compound	Low/High Water	Low/High Soil/Sediment
4-Bromofluorobenzene[a]	86 -115	74 - 121
Dibromofluoromethane[a]	86 - 118	80 - 120
Toluene-d$_8$[a]	88 - 110	81 - 117

[a] Single laboratory data for guidance only.

Method 8270A – *Table 2*

Estimated Quantitation Limits (EQL) for Semivolatile Organics[a]

Semivolatiles	CAS Number	Estimated Quantitation Limits[b]	
		Ground Water µg/L	Low Soil/ Sediment[1] µg/kg
Acenapthene	83-32-9	10	660
Acenaphthylene	208-96-8	10	660
Acetophenone	98-86-2	10	nd
2-Acetylaminofluorene	53-96-3	20	nd
1-Acetyl-2-thiourea	591-08-2	1000	nd
2-Aminoanthraquinone	117-79-3	20	nd
Aminoazobenzene	60-09-3	10	nd
4-Aminobiphenyl	92-67-1	20	nd
Anilazine	101-05-3	100	nd
o-Anisidine	90-04-0	10	nd
Anthracene	120-12-7	10	660
Aramite	140-57-8	20	nd
Azinphos-methyl	86-50-0	100	nd
Barban	101-27-9	200	nd
Benzo(a)anthracene	56-55-3	10	660
Benzo(b)fluoranthene	205-99-2	10	660
Benzo(k)fluoranthene	207-08-9	10	660
Benzoic acid	65-85-0	50	3300
Benzo(g,h,i)perylene	191-24-2	10	660
Benzo(a)pyrene	50-32-8	10	660
p-Benzoquinone	106-51-4	10	nd
Benzyl alcohol	100-51-6	20	1300
bis(2-Chloroethoxy) methane	111-91-1	10	660
bis(2-Chloroethyl) ether	111-44-4	10	660
bis(2-Chloroisopropyl) ether	108-60-1	10	660
4-Bromophenyl phenyl ether	101-55-3	10	660
Bromoxynil	1689-84-5	10	nd
Butyl benzyl phthalate	85-68-7	10	660
Captafol	2425-06-1	20	nd
Captan	133-06-2	50	nd
Carbaryl	63-25-2	10	nd
Carbofuran	1563-66-2	10	nd
Carbophenothion	786-19-6	10	nd
Chlorfenvinphos	470-90-6	20	nd
4-Chloroaniline	106-47-8	20	1300
Chlorobenzilate	510-15-6	10	nd
5-Chloro-2-methylaniline	95-79-4	10	nd
4-Chloro-3-methylphenol	59-50-7	20	1300
3-(Chloromethyl) pyridine hydrochloride	6959-48-4	100	nd

Estimated Quantitation Limits (EQL) for Semivolatile Organics[a] (Continued)

Semivolatiles	CAS Number	Estimated Quantitation Limits[b]	
		Ground Water μg/L	Low Soil/ Sediment[1] μg/kg
2-Chloronaphthalene	91-58-7	10	660
2-Chlorophenol	95-57-8	10	660
4-Chlorophenyl phenyl ether	7005-72-3	10	660
Chrysene	218-01-9	10	660
Coumaphos	56-72-4	40	nd
p-Cresidine	120-71-8	10	nd
Crotoxyphos	7700-17-6	20	nd
2-Cyclohexyl-4,6-dinitrophenol	131-89-5	100	nd
Demeton-o	298-03-3	10	nd
Demeton-s	126-75-0	10	nd
Diallate (*cis or trans*)	2303-16-4	10	nd
Diallate (*trans or cis*)	2303-16-4	10	nd
2,4-Diaminotoluene	95-80-7	20	nd
Dibenz(a,j)acridine	224-42-0	10	nd
Dibenz(a,h)anthracene	53-70-3	10	660
Dibenzofuran	132-64-9	10	660
Dibenzo(a,e)pyrene	192-65-4	10	nd
Di-n-butylphthalate	84-74-2	10	nd
Dichlone	117-80-6	na	nd
1,2-Dichlorobenzene	95-50-1	10	660
1,3-Dichlorobenzene	541-73-1	10	660
1,4-Dichlorobenzene	106-46-7	10	660
3,3'-Dichlorobenzidine	91-94-1	20	1300
2,4-Dichlorophenol	120-83-2	10	660
2,6-Dichlorophenol	84-65-0	10	nd
Dichlorovos	62-73-7	10	nd
Dicrotophos	141-66-2	10	nd
Diethylphthalate	84-66-2	10	660
Diethylstilbesterol	56-53-1	20	nd
Diethyl sulfate	64-67-5	100	nd
Dimethoate	60-51-5	20	nd
3,3'-Dimethoxybenzidine	119-90-4	100	nd
Dimethylaminoazobenzene	60-11-7	10	nd
7,12-Dimethylbenz(a)anthracene	57-97-6	10	nd
3,3'-Dimethylbenzidine	119-93-7	10	nd
α,α-Dimethylphenethylamine	122-09-8	nd	nd
2,4-Dimethylphenol	105-67-9	10	660
Dimethyl phthalate	99-65-0	10	660
1,2-Dinitrobenzene	131-11-3	40	nd
1,3-Dinitrobenzene	528-29-0	20	nd

Method 8270A – *Table 2*

Estimated Quantitation Limits (EQL) for Semivolatile Organics[a] (Continued)

Semivolatiles	CAS Number	Estimated Quantitation Limits[b]	
		Ground Water µg/L	Low Soil/ Sediment[1] µg/kg
1,4-Dinitrobenzene	100-25-4	40	nd
4,6-Dinitro-2-methylphenol	534-52-1	50	3300
2,4-Dinitrophenol	51-28-5	50	3300
2,4-Dinitrotoluene	121-14-2	10	660
2,6-Dinitrotoluene	606-20-2	10	660
Dinocap	39300-45-3	100	nd
Dinoseb	88-85-7	20	nd
5,5-Diphenylhydantoin	57-41-0	20	nd
Di-n-octyl phthalate	117-84-0	10	660
Disulfoton	298-04-4	10	nd
EPN	2104-64-5	10	nd
Ethion	563-12-2	10	nd
Ethyl carbamate	51-79-6	50	nd
bis(2-Ethylhexyl)phthalate	117-81-7	10	660
Ethyl methanesulfonate	62-50-0	20	nd
Famphur	52-85-7	20	nd
Fensulfothion	115-90-2	40	nd
Fenthion	55-38-9	10	nd
Fluchloralin	33245-39-5	20	nd
Fluoranthene	206-44-0	10	660
Fluorene	86-73-7	10	660
Hexachlorobenzene	118-74-1	10	660
Hexachlorobutadiene	87-68-3	10	660
Hexachlorocyclopentadiene	77-47-4	10	660
Hexachloroethane	67-72-1	10	660
Hexachlorophene	70-30-4	50	nd
Hexachloropropene	1888-71-7	10	nd
Hexamethyl phosphoramide	680-31-9	20	nd
Hydroquinone	123-31-9	nd	nd
Indeno(1,2,3-cd)pyrene	193-39-5	10	660
Isodrin	465-73-6	20	nd
Isophorone	78-59-1	10	660
Isosafrole	120-58-1	10	nd
Kepone	143-50-0	20	nd
Leptophos	21609-90-5	10	nd
Malathion	121-75-5	50	nd
Maleic anhydride	108-31-6	na	nd
Mestranol	72-33-3	20	nd
Methapyrilene	91-80-5	100	nd
Methoxychlor	72-43-5	10	nd

Method 8270A – *Table 2*

Estimated Quantitation Limits (EQL) for Semivolatile Organics[a] (Continued)

Semivolatiles	CAS Number	Estimated Quantitation Limits[b]	
		Ground Water µg/L	Low Soil/ Sediment[1] µg/kg
3-Methylcholanthrene	56-49-5	10	nd
4,4'-Methylenebis(2-chloroaniline)	101-14-4	na	nd
Methylmethanesulfonate	66-27-3	10	nd
2-Methylnaphthalene	91-57-6	10	660
Methyl parathion	298-00-0	10	nd
2-Methylphenol	95-48-7	10	660
3-Methylphenol	108-39-4	10	nd
4-Methylphenol	106-44-5	10	660
Mevinphos	7786-34-7	10	nd
Mexacarbate	315-18-4	20	nd
Mirex	2385-85-5	10	nd
Monocrotophos	6923-22-4	40	nd
Naled	300-76-5	20	nd
Naphthalene	91-20-3	10	660
1,4-Naphthoquinone	130-15-4	10	nd
1-Naphthylamine	134-32-7	10	nd
2-Naphthylamine	91-59-8	10	nd
Nicotine	54-11-5	20	nd
5-Nitroacenaphthene	602-87-9	10	nd
2-Nitroaniline	88-74-4	50	3300
3-Nitroaniline	99-09-2	50	3300
4-Nitroaniline	100-01-6	20	nd
5-Nitro-o-anisidine	99-59-2	10	nd
Nitrobenzene	98-95-3	10	660
4-Nitrobiphenyl	92-93-3	10	nd
Nitrofen	1836-75-5	20	nd
2-Nitrophenol	88-75-5	10	660
4-Nitrophenol	100-02-7	50	3300
5-Nitro-o-toluidine	99-55-8	10	nd
4-Nitroquinoline-1-oxide	56-57-5	40	nd
N-Nitrosodibutylamine	924-16-3	10	nd
N-Nitrosodiethylamine	55-18-5	20	nd
N-Nitrosodiphenylamine	86-30-6	10	660
N-Nitroso-di-n-propylamine	621-64-7	10	660
N-Nitrosopiperidine	100-75-4	20	nd
N-Nitrosopyrrolidine	930-55-2	40	nd
Octamethyl pyrophosphoramide	152-16-9	200	nd
4,4'-Oxydianiline	101-80-4	20	nd
Parathion	56-38-2	10	nd
Pentachlorobenzene	608-93-5	10	nd

Estimated Quantitation Limits (EQL) for Semivolatile Organics[a] (Continued)

Semivolatiles	CAS Number	Estimated Quantitation Limits[b]	
		Ground Water µg/L	Low Soil/ Sediment[1] µg/kg
Pentachloronitrobenzene	82-68-8	20	nd
Pentachlorophenol	87-86-5	50	3300
Phenacetin	62-44-2	20	nd
Phenanthrene	85-01-8	10	660
Phenobarbital	50-06-6	10	nd
Phenol	108-95-2	10	660
1,4-Phenylenediamine	106-50-3	10	nd
Phorate	298-02-2	10	nd
Phosalone	2310-17-0	100	nd
Phosmet	732-11-6	40	nd
Phosphamidon	13171-21-6	100	nd
Phthalic anhydride	85-44-9	100	nd
2-Picoline	109-06-8	nd	nd
Piperonyl sulfoxide	120-62-7	100	nd
Pronamide	23950-58-5	10	nd
Propylthiouracil	51-52-5	100	nd
Pyrene	129-00-0	10	660
Pyridine	110-86-1	nd	nd
Resorcinol	108-46-3	100	nd
Safrole	94-59-7	10	nd
Strychnine	57-24-9	40	nd
Sulfallate	95-06-7	10	nd
Terbufos	13071-79-9	20	nd
1,2,4,5-Tetrachlorobenzene	95-94-3	10	nd
2,3,4,6-Tetrachlorophenol	58-90-2	10	nd
Tetrachlorvinphos	961-11-5	20	nd
Tetraethyl pyrophosphate	107-49-3	40	nd
Thionazine	297-97-2	20	nd
Thiophenol (Benzenethiol)	108-98-5	20	nd
Toluene diisocyanate	584-84-9	100	nd
o-Toluidine	95-53-4	10	nd
1,2,4-Trichlorobenzene	120-82-1	10	660
2,4,5-Trichlorophenol	95-95-4	10	660
2,4,6-Trichlorophenol	88-06-2	10	660
Trifluralin	1582-09-8	10	nd
2,4,5-Trimethylaniline	137-17-7	10	nd
Trimethyl phosphate	512-56-1	10	nd
1,3,5-Trinitrobenzene	99-35-4	10	nd
Tris(2,3-dibromopropyl) phosphate	126-72-7	200	nd

Method 8270A – *Table 2*

Estimated Quantitation Limits (EQL) for Semivolatile Organics[a] (Continued)

| Semivolatiles | CAS Number | Estimated Quantitation Limits[b] | |
		Ground Water µg/L	Low Soil/ Sediment[1] µg/kg
Tri-p-tolyl phosphate(h)	78-32-0	10	nd
0,0,0-Triethylphosphorothioate	126-68-1	nt	nd

[a] EQLs listed for soil/sediment are based on wet weight. Normally data is reported on a dry weight basis, therefore, EQLs will be higher based on the % dry weight of each sample. This is based on a 30-g sample and gel permeation chromatography cleanup.

[b] Sample EQLs are highly matrix-dependent. The EQLs listed herein are provided for guidance and may not always be achievable.

nd = Not determined
na = Not applicable
nt = Not tested

Other Matrice	Factor[1]
Medium-level soil and sludges by sonicator	7.5
Non-water miscible waste	75

[1] EQL - [EQL for Low Soil/Sediment (Table2)] X [Factor]

Method 8270A – *Table 3*

DFTPP KEY Ions and Ion Abundance Criteria[a]

Mass	Ion Abundance Criteria
51	30-60% of mass 198
68	<2% of mass 69
70	<2% of mass 69
127	40-60% of mass 198
197	<1% of mass 198
198	Base peak, 100% relative abundance
199	5-9% of mass 198
275	10-30% of mass 198
365	>1% of mass 198
441	Present but less than mass 443
442	>40% of mass 198
443	17-23% of mass 442

[a] J.W. Eichelberger, L.E. Harris, and W.L. Budde. Reference Compound to Calibrate Ion Abundance Measurement in Gas Chromatograpy-Mass Spectrometry, Analytical Chemistry, 47, 995 (1975).

Method 8270A – *Table 6*

QC Acceptance Criteria[a]

Parameter	Test conc. (µg/L)	Limit for s (µg/L)	Range for $\bar{x}$ (µg/L)	Range for P, P_s (%)
Acenaphthene	100	27.6	60.1-132.3	47-145
Acenaphthylene	100	40.2	53.5-126.0	33-145
Aldrin	100	39.0	7.2-152.2	D-166
Anthracene	100	32.0	43.4-118.0	27-133
Benzo(a)anthracene	100	27.6	41.8-133.0	33-143
Benzo(b)fluoranthene	100	38.8	42.0-140.4	24-159
Benzo(k)fluoranthene	100	32.3	25.2-145.7	11-162
Benzo(a)pyrene	100	39.0	31.7-148.0	17-163
Benzo(g,h,i)perylene	100	58.9	D-195.0	D-219
Benzyl butyl phthalate	100	23.4	D-139.9	D-152
β-BHC	100	31.5	41.5-130.6	24-149
δ-BHC	100	21.6	D-100.0	D-110
bis(2-Chloroethyl)ether	100	55.0	42.9-126.0	12-158
bis(2-Chloroethoxy)methane	100	34.5	49.2-164.7	33-184
bis(2-Chloroisopropyl)ether	100	46.3	62.8-138.6	36-166
bis(2-Ethylhexyl)phthalate	100	41.1	28.9-136.8	8-158
4-Bromophenyl phenyl ether	100	23.0	64.9-114.4	53-127
2-Chloronaphthalene	100	13.0	64.5-113.5	60-118
4-Chlorophenyl phenyl ether	100	33.4	38.4-144.7	25-158
Chrysene	100	48.3	44.1-139.9	17-168
4,4'-DDD	100	31.0	D-134.5	D-145
4,4'-DDE	100	32.0	19.2-119.7	4-136
4,4'-DDT	100	61.6	D-170.6	D-203
Dibenzo(a,h)anthracene	100	70.0	D-199.7	D-227
Di-n-butyl phthalate	100	16.7	8.4-111.0	1-118
1,2-Dichlorobenzene	100	30.9	48.6-112.0	32-129
1,3-Dichlorobenzene	100	41.7	16.7-153.9	D-127
1,4-Dichlorobenzene	100	32.1	37.3-105.7	20-124
3,3'-Dichlorobenzidine	100	71.4	8.2-212.5	D-262
Dieldrin	100	30.7	44.3-119.3	29-136
Diethyl phthalate	100	26.5	D-100.0	D-114
Dimethyl phthalate	100	23.2	D-100.0	D-112
2,4-Dinitrotoluene	100	21.8	47.5-126.9	39-139
2,6-Dinitrotoluene	100	29.6	68.1-136.7	50-158

Method 8270A – *Table 6*

QC Acceptance Criteria[a] (Continued)

Parameter	Test conc. (μg/L)	Limit for s (μg/L)	Range for $\bar{x}$ (μg/L)	Range for P, P_s (%)
Di-n-octylphthalate	100	31.4	18.6-131.8	4-146
Endosulfan sulfate	100	16.7	D-103.5	D-107
Endrin aldehyde	100	32.5	D-188.8	D-209
Fluoranthene	100	32.8	42.9-121.3	26-137
Fluorene	100	20.7	71.6-108.4	59-121
Heptachlor	100	37.2	D-172.2	D-192
Heptachlor epoxide	100	54.7	70.9-109.4	26-155
Hexachlorobenzene	100	24.9	7.8-141.5	D-152
Hexachlorobutadiene	100	26.3	37.8-102.2	24-116
Hexachloroethane	100	24.5	55.2-100.0	40-113
Indeno(1,2,3-cd)pyrene	100	44.6	D-150.9	D-171
Isophorone	100	63.3	46.6-180.2	21-196
Naphthalene	100	30.1	35.6-119.6	21-133
Nitrobenzene	100	39.3	54.3-157.6	35-180
N-Nitrosodi-n-propylamine	100	55.4	13.6-197.9	D-230
PCB-1260	100	54.2	19.3-121.0	D-164
Phenanthrene	100	20.6	65.2-108.7	54-120
Pyrene	100	25.2	69.6-100.0	52-115
1,2,4-Trichlorobenzene	100	28.1	57.3-129.2	44-142
4-Chloro-3-methylphenol	100	37.2	40.8-127.9	22-147
2-Chlorophenol	100	28.7	36.2-120.4	23-134
2,4-Chlorophenol	100	26.4	52.5-121.7	39-135
2,4-Dimethylphenol	100	26.1	41.8-109.0	32-119
2,4-Dinitrophenol	100	49.8	D-172.9	D-191
2-Methyl-4,6-dinitrophenol	100	93.2	53.0-100.0	D-181
2-Nitrophenol	100	35.2	45.0-166.7	29-182
4-Nitrophenol	100	47.2	13.0-106.5	D-132
Pentachlorophenol	100	48.9	38.1-151.8	14-176
Phenol	100	22.6	16.6-100.0	5-112
2,4,6-Trichlorophenol	100	31.7	52.4-129.2	37-144

s   = Standard deviation of four recovery measurements, in μg/L

$\bar{x}$ = Average recovery for four recovery measurements, in μg/L

P, P_s = Percent recovery measured

D   = Detected; result must be greater than zero.

[a]   Criteria from 40 CFR Part 136 for Method 625. These criteria are based directly on the method performance data in Table 7. Where necessary, the limits for recovery have been broadened to assure applicability of the limits to concentrations below those used to develop Table 7.

Method 8270A – *Table 8*

Surrogate Spike Recovery Limits For Water and Soil/Sediment Samples

Surrogate Compound	Low/High Water	Low/High Soil/Sediment
Nitrobenzene-d$_5$	35-114	23-120
2-Fluorobiphenyl	43-116	30-115
p-Terphenyl-d$_{14}$	33-141	18-137
Phenol-d$_6$	10-94	24-113
2-Fluorophenol	21-100	25-121
2,4,6-Tribromophenol	10-123	19-122

APPENDIX B

All organic analytes mentioned in any of the test methods covered in this reference are listed in alphabetical order followed by their CAS number. If the analyte is included in one of the listed test methods it is noted with a "•" in the appropriate column.

Organic Analyte Cross-Reference to Methods

Analyte	CAS Number	Herbicides			Pesticides & PCBs					Semivolatile Organic					Volatile Organic							Aromatic Volatile				Halogenated Volatile					
		SW846 - 8150A	EPA 500 - 515.1	Std. Methods 6640	SW846 - 8080	EPA500 - 508	EPA600 - 608	Std. Methods 6630	CLP-PEST- Organic	SW846 - 8270A	EPA500 - 525.1	EPA600 - 625	Std. Methods 6410	CLP - SVOA	SW846 - 8240A	SW846 - 8260	EPA500 - 524.1	EPA500 - 524.2	EPA600 - 624	Std. Methods 6210	CLP - VOA	SW-846 - 8020	EPA500 - 503.1	EPA600 - 602	Std. Methods 6220	SW846 - 8010A	SW-846 - 8021	EPA500 - 502.1	EPA500 - 502.2	EPA600 - 601	Std. Methods 6230
Acenaphthene	83-32-9									•		•	•	•																	
Acenaphthylene	208-96-8									•	•	•	•	•																	
Acetone	67-64-1														•					•											
Acetonitrile	75-05-8														•																
Acetophenone	98-86-2									•																					
1-Acetyl-2-thiourea	591-08-2									•																					
2-Acetylaminofluorene	53-96-3									•																					
Acifluorfen	50594-66-6	•																													
Acrolein	107-02-8														•																
Acrylonitrile	107-13-1														•																
Alachlor	15972-60-8										•																				
Aldrin	309-00-2				•	•	•	•	•	•	•	•	•																		
Allyl alcohol	107-18-6														•																
Allyl chloride	107-05-1														•																
2-Aminoanthraquinone	117-79-3									•																					
Aminoazobenzene	60-09-3									•																					
4-Aminobiphenyl	92-67-1									•																					
Anilazine	101-05-3									•																					
Aniline	62-53-3									•																					
o-Anisidine	90-04-0									•																					
Anthracene	120-12-7									•	•	•	•	•																	
Aramite	140-57-8									•																					
Aroclor-1016	12674-11-2				•	•	•	•	•	•		•	•																		
Aroclor-1221	11104-28-2				•	•	•	•	•	•		•	•																		
Aroclor-1232	11141-16-5				•	•	•	•	•	•		•	•																		
Aroclor-1242	53469-21-9				•	•	•	•	•	•		•	•																		
Aroclor-1248	12672-29-6				•	•	•	•	•	•		•	•																		
Aroclor-1254	11097-69-1				•	•	•	•	•	•		•	•																		
Aroclor-1260	11096-82-5				•	•	•	•	•	•		•	•																		
Atrazine	1912-24-9										•																				
Azinphos methyl	86-50-0									•																					
Barban	101-27-9									•																					
Bentazon	25057-89-0	•																													
Benzene	71-43-2														•	•	•	•	•	•	•	•	•	•	•		•		•		
Benzidine	92-87-5									•		•	•																		
Benzoic acid	65-85-0									•																					
p-Benzoquinone	106-51-4									•																					
Benzo(a)anthracene	56-55-3									•	•	•	•	•																	
Benzo(a)pyrene	50-32-8									•	•	•	•	•																	
Benzo(b)fluoranthene	205-99-2									•	•	•	•	•																	
Benzo(g,h,i)perylene	191-24-2									•	•	•	•	•																	
Benzo(k)fluoranthene	207-08-9									•	•	•	•	•																	
Benzyl alcohol	100-51-6									•																					

Organic Analyte Cross-Reference to Methods

Method column groups:
- **Herbicides:** SW846-8150A, EPA500-515.1, Std. Methods 6640
- **Pesticides & PCBs:** SW846-8080, EPA500-508, EPA600-608, Std. Methods 6630, CLP-PEST-Organic
- **Semivolatile Organic:** SW846-8270A, EPA500-525.1, EPA600-625, Std. Methods 6410, CLP-SVOA
- **Volatile Organic:** SW846-8240A, SW846-8260, EPA500-524.1, EPA500-524.2, EPA600-624, Std. Methods 6210, CLP-VOA
- **Aromatic Volatile:** SW846-8020, EPA500-503.1, EPA600-602, Std. Methods 6220
- **Halogenated Volatile:** SW846-8010A, SW-846-8021, EPA500-502.1, EPA500-502.2, EPA600-601, Std. Methods 6230

Analyte	CAS Number	8150A	515.1	6640	8080	508	608	6630	CLP-PEST	8270A	525.1	625	6410	CLP-SVOA	8240A	8260	524.1	524.2	624	6210	CLP-VOA	8020	503.1	602	6220	8010A	8021	502.1	502.2	601	6230
Benzyl butyl phthalate	85-68-7											•	•																		
Benzyl chloride (alpha-chlorotoluene)	100-44-7														•											•					
alpha-BHC	319-84-6				•	•	•	•	•	•		•	•																		
beta-BHC	319-85-7				•	•	•	•	•	•		•	•																		
delta-BHC	319-86-8				•	•	•	•	•	•		•	•																		
gamma-BHC (Lindane)	58-89-9				•	•	•	•	•	•	•	•	•																		
Bromoacetone	598-31-2														•																
Bromobenzene	108-86-1															•	•	•		•		•				•	•	•	•		
Bromochloromethane	74-97-5															•	•	•		•						•	•		•		
Bromodichloromethane	75-27-4														•	•	•	•	•	•						•	•	•	•	•	•
Bromoform	75-25-2														•	•	•	•	•	•						•	•	•	•	•	•
Bromomethane	74-83-9														•	•	•	•	•	•						•	•	•	•	•	•
4-Bromophenyl-phenylether	101-55-3									•		•	•	•																	
Bromoxynil	1689-84-5									•																					
2-Butanone	78-93-3														•						•										
Butylbenzylphthalate	85-68-7									•	•		•																		
n-Butylbenzene	104-51-8															•	•	•				•					•		•		
sec-Butylbenzene	135-98-8															•	•	•				•					•		•		
tert-Butylbenzene	98-06-6															•	•	•				•					•		•		
Captafol	2425-06-1									•																					
Captan	133-06-2							•		•																					
Carbaryl	63-25-2									•																					
Carbazole	86-74-8													•																	
Carbofuran	1563-66-2									•																					
Carbon disulfide	75-15-0														•						•										
Carbon tetrachloride	56-23-5														•	•	•	•	•	•						•	•	•	•	•	•
Carbophenothion	786-19-6									•																					
Chloramben	133-90-4	•																													
alpha-Chlordane	5103-71-9					•			•	•																					
Chlordane	57-74-9				•	•	•	•	•	•		•	•																		
gamma-Chlordane	5103-74-2					•			•	•																					
trans-Nonachlor-chlordane	39765-80-5									•																					
Chlorfenvinphos	470-90-6									•																					
Chlorneb	2675-77-6					•																									
5-Chloro-2-methylaniline	95-79-4									•																					
4-Chloro-3-methylphenol	59-50-7									•		•	•	•																	
4-Chloroaniline	106-47-8									•				•																	
Chlorobenzene	108-90-7														•	•	•	•	•	•	•	•	•	•	•	•	•	•	•	•	•
Chlorobenzilate	510-15-6					•				•																					
2-Chlorobiphenyl	2051-60-7										•																				
Chloroethane	75-00-3														•	•	•	•	•	•	•					•	•	•	•	•	•
2-Chloroethanol	107-07-3														•																

Organic Analyte Cross-Reference to Methods

Analyte	CAS Number	Herbi-cides			Pesticides & PCBs					Semivolatile Organic					Volatile Organic							Aromatic Volatile				Halogenated Volatile					
		SW846 - 8150A	EPA 500 - 515.1	Std. Methods 6640	SW846 - 8080	EPA500 - 508	EPA600 - 608	Std. Methods 6630	CLP-PEST-Organic	SW846 - 8270A	EPA500 - 525.1	EPA600 - 625	Std. Methods 6410	CLP - SVOA	SW846 - 8240A	SW846 - 8260	EPA500 - 524.1	EPA500 - 524.2	EPA600 - 624	Std. Methods 6210	CLP - VOA	SW-846 - 8020	EPA500 - 503.1	EPA600 - 602	Std. Methods 6220	SW846 - 8010A	SW-846 - 8021	EPA500 - 502.1	EPA500 - 502.2	EPA600 - 601	Std. Methods 6230
bis(2-Chloroethoxy)methane	111-91-1									•		•	•	•																	
2-Chloroethylvinyl ether	110-75-8														•				•	•						•				•	•
bis(2-Chloroethyl)ether	111-44-4									•		•	•	•																	
Chloroform	67-66-3														•	•	•	•	•	•	•					•	•	•	•	•	•
bis(2-Chloroisopropyl)ether	108-60-1									•		•	•	•																	
Chloromethane	74-87-3														•	•	•	•	•	•	•					•	•	•	•	•	•
3-(Chloromethyl) pyridine hydrochloride	6959-48-4									•																					
1-Chloronaphthalene	90-13-1									•																					
2-Chloronaphthalene	91-58-7									•		•	•	•																	
2-Chlorophenol	95-57-8									•		•	•	•																	
4-Chlorophenyl-phenylether	7005-72-3									•		•	•	•																	
Chloroprene	126-99-8														•																
3-Chloropropionitrile	542-76-7														•																
Chlorothalonil	2921-88-2				•																										
2-Chlorotoluene	95-49-8														•	•		•		•		•					•		•		•
4-Chlorotoluene	106-43-4														•	•		•		•		•					•		•		•
Chrysene	218-01-9									•	•	•	•	•																	
Coumaphos	56-72-4									•																					
p-Cresidine	120-71-8									•																					
Crotoxyphos	7700-17-6									•																					
2-Cyclohexyl-4,6-dinitrophenol	131-89-5									•																					
2,4'-D	94-75-7	•	•	•																											
Dalapon	75-99-0	•	•																												
2,4-DB	94-82-6	•	•																												
DCPA	1897-45-6				•																										
DCPA acid metabolites	N/A	•																													
4,4'-DDD	72-54-8				•	•	•	•	•	•		•	•																		
4,4'-DDE	72-55-9				•	•	•	•	•	•		•	•																		
4,4'-DDT	50-29-3				•	•	•	•	•	•		•	•																		
Demeton-o	298-03-3									•																					
Demeton-s	126-75-0									•																					
Diallate (trans or cis)	2303-16-4									•																					
2,4-Diaminotoluene	95-80-7									•																					
Dibenzofuran	132-64-9									•				•																	
Dibenzo(a,e)pyrene	192-65-4									•																					
Dibenz(a,h)anthracene	53-70-3									•	•	•	•	•																	
Dibenz(a,j)acridine	224-42-0									•																					
1,2-Dibromo-3-chloropropane	96-12-8														•	•		•	•	•									•		•
Dibromochloromethane	124-48-1														•	•	•	•	•	•	•					•	•	•	•	•	•
1,2-Dibromoethane	106-93-4														•	•		•	•	•									•	•	•
Dibromomethane	74-95-3														•	•		•		•						•	•		•		•
Dicamba	1918-00-9	•	•																												

Organic Analyte Cross-Reference to Methods

Analyte	CAS Number	Herbicides			Pesticides & PCBs					Semivolatile Organic					Volatile Organic							Aromatic Volatile				Halogenated Volatile					
		SW846-8150A	EPA500-515.1	Std. Methods 6640	SW846-8080	EPA500-508	EPA600-608	Std. Methods 6630	CLP-PEST-Organic	SW846-8270A	EPA500-525.1	EPA600-625	Std. Methods 6410	CLP-SVOA	SW846-8240A	SW846-8260	EPA500-524.1	EPA500-524.2	EPA600-624	Std. Methods 6210	CLP-VOA	SW-846-8020	EPA500-503.1	EPA600-602	Std. Methods 6220	SW846-8010A	SW-846-8021	EPA500-502.1	EPA500-502.2	EPA600-601	Std. Methods 6230
Dichlone	117-80-6									●																					
Dichloran	N/A							●																							
1,4-Dichloro-2-butene	764-41-0														●																
1,2-Dichlorobenzene	95-50-1									●		●	●	●	●	●		●	●	●		●	●	●	●	●	●	●	●	●	●
1,3-Dichlorobenzene	541-73-1									●		●	●	●	●	●		●		●		●	●	●	●	●	●	●	●	●	●
1,4-Dichlorobenzene	106-46-7									●		●	●	●	●	●		●	●	●		●	●	●	●	●	●	●	●	●	●
3,3'-Dichlorobenzidine	91-94-1									●			●	●																	
3,5-Dichlorobenzoic acid	51-36-5	●																													
2,3-Dichlorobiphenyl	16605-91-7									●																					
Dichlorodifluoromethane	75-71-8														●	●		●		●						●	●	●	●	●	
1,1-Dichloroethane	75-34-3														●	●	●	●		●	●					●	●	●	●	●	●
1,2-Dichloroethane	107-06-2														●	●	●	●		●	●					●	●	●	●	●	●
1,1-Dichloroethene	75-35-4														●	●	●	●		●	●					●	●	●	●	●	●
cis-1,2-Dichloroethene	156-59-4														●	●		●		●						●	●		●		
trans-1,2-Dichloroethene	156-60-5														●	●	●	●		●	●					●	●		●		
1,2-Dichloroethene (total)	540-59-0																					●										
2,4-Dichlorophenol	120-83-2									●		●	●	●																	
2,6-Dichlorophenol	87-65-0									●																					
Dichloroprop	120-36-5	●	●																												
1,2-Dichloropropane	78-87-5														●	●	●	●	●	●	●					●	●	●	●	●	●
1,3-Dichloropropane	142-28-9														●	●		●		●						●	●		●		
2,2-Dichloropropane	590-20-7														●	●		●		●						●	●		●		
1,3-Dichloro-2-propanol	96-23-1														●																
1,1-Dichloropropene	563-58-6														●	●		●		●						●	●		●		
cis-1,3-Dichloropropene	10061-01-5														●	●	●	●		●	●					●	●	●	●	●	●
trans-1,3-Dichloropropene	10061-02-6														●	●	●	●		●	●					●	●	●	●	●	●
Dichlorovos	62-73-7									●																					
Dicrotophos	141-66-2									●																					
Dieldrin	60-57-1				●	●	●	●	●	●			●	●																	
1,2,3,4-Diepoxybutane	298-18-0														●																
Diethyl sulfate	64-67-5									●																					
Diethylphthalate	84-66-2									●	●	●	●	●																	
Diethylstilbesterol	56-53-1									●																					
Dimethoate	60-51-5									●																					
3,3'-Dimethoxybenzidine	119-90-4									●																					
p-Dimethylaminoazobenzene	60-11-7									●																					
3,3'-Dimethylbenzidine	119-93-7									●																					
7,12-Dimethylbenz[A]anthracene	57-97-6									●																					
2,4-Dimethylphenol	105-67-9									●		●	●	●																	
Dimethylphthalate	131-11-3									●	●	●	●	●																	
alpha, alpha-Dimethyphenethylamine	122-09-8									●																					

Organic Analyte Cross-Reference to Methods

Analyte	CAS Number	Herbicides			Pesticides & PCBs					Semivolatile Organic					Volatile Organic							Aromatic Volatile				Halogenated Volatile					
		SW846 - 8150A	EPA 500 - 515.1	Std. Methods 6640	SW846 - 8080	EPA500 - 508	EPA600 - 608	Std. Methods 6630	CLP-PEST-Organic	SW846 - 8270A	EPA500 - 525.1	EPA600 - 625	Std. Methods 6410	CLP - SVOA	SW846 - 8240A	SW846 - 8260	EPA500 - 524.1	EPA500 - 524.2	EPA600 - 624	Std. Methods 6210	CLP - VOA	SW-846 - 8020	EPA500 - 503.1	EPA600 - 602	Std. Methods 6220	SW846 - 8010A	SW-846 - 8021	EPA500 - 502.1	EPA500 - 502.2	EPA600 - 601	Std. Methods 6230
Di-n-butylphthalate	84-74-2									•	•	•	•	•																	
4,6-Dinitro-2-methylphenol	121-14-2									•				•																	
1,3-Dinitrobenzene	99-65-0									•																					
1,4-Dinitrobenzene	100-25-4									•																					
1,2-Dinitrobenzene	528-29-0									•																					
2,4-Dinitrophenol	51-28-5									•		•	•	•																	
2,4-Dinitrotoluene	121-14-2									•		•	•	•																	
2,6-Dinitrotoluene	606-20-2									•		•	•	•																	
Dinocap	6119-92-2																														
Di-n-octylphthalate	117-84-0									•		•	•	•																	
Dinoseb	88-85-7	•	•							•																					
1,4 Dioxane	123-91-1														•																
Diphenylamine	122-39-4									•																					
5,5-Diphenylhydantoin	57-41-0									•																					
1,2 Diphenylhydrazine	122-66-7									•																					
Disulfoton	298-04-4									•																					
Di(2-ethylhexyl)adipate	103-23-1										•																				
Endosulfan I	959-98-8				•	•	•	•	•	•		•	•																		
Endosulfan II	33213-65-9				•	•	•	•	•	•		•	•																		
Endosulfan sulfate	1031-07-8				•	•	•	•	•	•		•	•																		
Endrin	72-20-8				•	•	•	•	•	•	•	•	•																		
Endrin aldehyde	7421-93-4				•	•	•	•	•	•		•	•																		
Endrin ketone	53494-70-5								•	•																					
Epichlorohydrin	106-89-8														•																
EPN	2104-64-5									•																					
Ethanol	64-17-5														•																
Ethion	563-12-2									•																					
Ethyl benzene	100-41-4														•	•	•	•	•	•	•	•	•	•	•	•			•		•
Ethyl carbamate	51-79-6									•																					
Ethyl methacrylate	97-63-2														•																
Ethyl methanesulfonate	62-50-0									•																					
Ethylene oxide	75-21-8														•																
bis(2-Ethylhexyl)phthalate	117-81-7									•	•	•	•	•																	
Etridiazole	2593-15-9					•																									
Famphur	52-85-7									•																					
Fensulfothion	115-90-2									•																					
Fenthion	55-38-9									•																					
Fluchloralin	33245-39-5									•																					
Fluoranthene	206-44-0									•		•	•	•																	
Fluorene	86-73-7									•	•	•	•	•																	
Heptachlor	76-44-8				•	•	•	•	•	•	•	•	•																		
Heptachlor epoxide	1024-57-3				•	•	•	•	•	•	•	•	•																		

Organic Analyte Cross-Reference to Methods

Analyte	CAS Number	Herbi-cides			Pesticides & PCBs					Semivolatile Organic					Volatile Organic							Aromatic Volatile				Halogenated Volatile					
		SW846-8150A	EPA 500-515.1	Std. Methods 6640	SW846-8080	EPA500-508	EPA600-608	Std. Methods 6630	CLP-PEST-Organic	SW846-8270A	EPA500-525.1	EPA600-625	Std. Methods 6410	CLP-SVOA	SW846-8240A	SW846-8260	EPA500-524.1	EPA500-524.2	EPA600-624	Std. Methods 6210	CLP-VOA	SW-846-8020	EPA500-503.1	EPA600-602	Std. Methods 6220	SW846-8010A	SW-846-8021	EPA500-502.1	EPA500-502.2	EPA600-601	Std. Methods 6230
2,2',3,3',4,4',6-Heptachlorobiphenyl	52663-71-5										•																				
Hexachlorobenzene	118-74-1					•				•	•	•	•	•																	
2,2',4,4',5,6'-Hexachlorobiphenyl	60145-22-4										•																				
Hexachlorobutadiene	87-68-3									•		•	•	•	•		•			•				•						•	•
Hexachlorocyclopentadiene	77-47-4									•	•	•	•	•																	
Hexachloroethane	67-72-1									•		•	•	•																	
Hexachlorophene	70-30-4									•																					
Hexachloropropene	1888-71-7									•																					
Hexamethyl phosphoramide	680-31-9									•																					
2-Hexanone	591-78-6														•						•										
Hydroquinone	123-31-9									•																					
5-Hydroxydicamba	7600-50-2	•																													
2-Hydroxypropionitrile	78-97-7														•																
Indeno(1,2,3-cd)pyrene	193-39-5									•	•	•	•	•																	
Iodomethane	74-88-4														•																
Isobutyl alcohol	78-83-1														•																
Isodrin	465-73-6									•																					
Isophorone	78-59-1									•		•	•	•																	
Isopropylbenzene	98-82-8															•		•		•		•						•	•		
4-Isopropyltoluene	99-87-6															•		•		•		•						•	•		
Isosafrole	120-58-1									•																					
Kepone	143-50-0									•																					
Leptophos	21609-90-5									•																					
Maleic anhydride	108-31-6									•																					
Malononitrile	109-77-3														•																
MCPA	94-74-6	•																													
MCPP	93-65-2	•																													
Malathion	121-75-5							•		•																					
Mevinphos	7786-34-7									•																					
Mestranol	72-33-3									•																					
Methacrylonitrile	126-98-7														•																
Methapyrilene	91-80-5									•																					
Methoxychlor	72-43-5				•	•		•	•	•	•																				
Methyl iodide	74-88-4														•																
Methyl methacrylate	80-62-6														•																
Methyl parathion	298-00-0							•		•																					
4-Methyl-2-pentanone	108-10-1														•						•										
2-Methyl-4,6-dinitrophenol	534-52-1												•	•																	
3-Methylcholanthrene	56-49-5									•																					
Methylene chloride	75-09-2														•	•	•	•	•	•	•					•	•	•	•	•	•

Organic Analyte Cross-Reference to Methods

Analyte	CAS Number	Herbicides SW846-8150A	EPA 500-515.1	Std. Methods 6640	Pesticides & PCBs SW846-8080	EPA500-508	EPA600-608	Std. Methods 6630	CLP-PEST-Organic	Semivolatile SW846-8270A	EPA500-525.1	EPA600-625	Std. Methods 6410	CLP-SVOA	Volatile SW846-8240A	SW846-8260	EPA500-524.1	EPA500-524.2	EPA600-624	Std. Methods 6210	CLP-VOA	Aromatic SW-846-8020	EPA500-503.1	EPA600-602	Std. Methods 6220	Halogenated SW846-8010A	SW-846-8021	EPA500-502.1	EPA500-502.2	EPA600-601	Std. Methods 6230
4,4'-Methylenebis(2-chloroaniline)	101-14-4									•																					
Methylmethanesulfonate	66-27-3									•																					
2-Methylnaphthalene	91-57-6									•				•																	
2-Methylphenol	95-48-7									•																					
3-Methylphenol	108-39-4									•																					
4-Methylphenol	106-44-5									•				•																	
Mexacarbate	315-18-4									•																					
Mirex	2385-85-5								•	•																					
Monocrotophos	6923-22-4									•																					
Naled	300-76-5									•																					
Naphthalene	91-20-3									•		•	•	•	•			•	•		•	•					•		•		
1,4-Naphthoquinone	130-15-4									•																					
2-Naphthylamine	91-59-8									•																					
1-Naphthylamine	134-32-7									•																					
Nicotine	54-11-5									•																					
5-Nitroacenaphthene	602-87-9									•																					
2-Nitroaniline	88-74-4									•				•																	
3-Nitroaniline	99-09-2									•				•																	
4-Nitroaniline	100-01-6									•				•																	
Nitrobenzene	98-95-3									•		•	•	•																	
4-Nitrobiphenyl	92-93-3									•																					
Nitrofen	1836-75-5									•																					
5-Nitro-o-anisidine	99-59-2									•																					
5-Nitro-o-toluidine	99-55-8									•																					
2-Nitrophenol	88-75-5									•		•	•	•																	
4-Nitrophenol	100-02-7	•								•		•	•	•																	
Nitroquinoline-1-oxide	56-57-5									•																					
N-Nitrosodibutylamine	924-16-3									•																					
N-Nitrosodiethylamine	55-18-5									•																					
N-Nitroso-di-n-propylamine	621-64-7									•		•	•	•																	
N-Nitrosodiphenylamine	86-30-6									•		•	•	•																	
N-Nitrosomethylethylamine	10595-95-6									•																					
N-Nitrosodimethylamine	62-75-9									•		•	•																		
N-Nitrosomorpholine	59-89-2									•																					
N-Nitrosopiperidine	100-75-4									•																					
N-Nitrosopyrrolidine	930-55-2									•																					
2,2',3,3',4,5',6,6'-Octachlorobiphenyl	40186-71-8										•																				
Octamethyl pyrophosphoramide	152-16-9									•																					
4,4'-Oxydianiline	101-80-4									•																					
Parathion	56-38-2								•	•																					

Organic Analyte Cross-Reference to Methods

Analyte	CAS Number	Herbicides			Pesticides & PCBs					Semivolatile Organic					Volatile Organic							Aromatic Volatile				Halogenated Volatile					
		SW846-8150A	EPA 500-515.1	Std. Methods 6640	SW846-8080	EPA500-508	EPA600-608	Std. Methods 6630	CLP-PEST-Organic	SW846-8270A	EPA500-525.1	EPA600-625	Std. Methods 6410	CLP-SVOA	SW846-8240A	SW846-8260	EPA500-524.1	EPA500-524.2	EPA600-624	Std. Methods 6210	CLP-VOA	SW-846-8020	EPA500-503.1	EPA600-602	Std. Methods 6220	SW846-8010A	SW-846-8021	EPA500-502.1	EPA500-502.2	EPA600-601	Std. Methods 6230
Pentachlorobenzene	608-93-5									●																					
2,2',3',4,6-Pentachlorobiphenyl	60233-25-2										●																				
Pentachloroethane	76-01-7														●																
Pentachloronitrobenzene (PCNB)	82-68-8							●		●																					
Pentachlorophenol	87-86-5	●								●	●	●	●	●																	
cis-Permethrin	52645-53-1					●																									
trans-Permethrin	52645-53-1					●																									
Phenacetin	62-44-2									●																					
Phenanthrene	85-01-8									●	●	●	●	●																	
Phenobarbital	50-06-6									●																					
Phenol	108-95-2									●		●	●	●																	
1,4-Phenylenediamine	106-50-3									●																					
Phorate	298-02-2									●																					
Phosalone	2310-17-0									●																					
Phosmet	732-11-6									●																					
Phosphamidon	13171-21-6									●																					
Phthalic anhydride	85-44-9									●																					
Picloram	1918-02-1		●																												
2-Picoline	109-06-8									●					●																
Piperonyl sulfoxide	120-62-7									●																					
Pronamide	23950-58-5									●																					
Propachlor	1918-16-7					●																									
Propargyl alcohol	107-19-7														●																
b-Propiolactone	57-57-8														●																
Propionitrile	107-12-0														●																
n-Propylamine	107-10-8														●																
n-Propylbenzene	103-65-1															●		●		●		●					●		●		
Propylthiouracil	51-52-5									●																					
Pyrene	129-00-0									●	●	●	●	●																	
Pyridine	110-86-1									●					●																
Resorcinol	108-46-3									●																					
Safrole	94-59-7									●																					
Silvex	93-72-1			●																											
Simazine	122-34-9										●																				
Strobane	71-55-6							●																							
Strychnine	57-24-9									●																					
Styrene	100-42-5														●	●	●	●		●	●		●				●		●		
Sulfallate	95-06-7									●																					
2,4,5-T	93-76-5	●	●	●																											
Terbufos	13071-79-9									●																					
1,2,4,5-Tetrachlorobenzene	95-94-3									●																					
2,2',4,4'-Tetrachlorobiphenyl	2437-79-8										●																				

Organic Analyte Cross-Reference to Methods

Analyte	CAS Number	Herbi-cides			Pesticides & PCBs					Semivolatile Organic					Volatile Organic							Aromatic Volatile				Halogenated Volatile					
		SW846 - 8150A	EPA 500 - 515.1	Std. Methods 6640	SW846 - 8080	EPA500 - 508	EPA600 - 608	Std. Methods 6630	CLP-PEST - Organic	SW846 - 8270A	EPA500 - 525.1	EPA600 - 625	Std. Methods 6410	CLP SVOA	SW846 - 8240A	SW846 - 8260	EPA500 - 524.1	EPA500 - 524.2	EPA600 - 624	Std. Methods 6210	CLP - VOA	SW-846 - 8020	EPA500 - 503.1	EPA600 - 602	Std. Methods 6220	SW846 - 8010A	SW-846 - 8021	EPA500 - 502.1	EPA500 - 502.2	EPA600 - 601	Std. Methods 6230
1,1,1,2-Tetrachloroethane	630-20-6														•		•	•		•						•	•	•	•		
1,1,2,2-Tetrachloroethane	79-34-5														•		•	•	•	•	•					•	•	•	•	•	•
Tetrachloroethene	127-18-4														•		•	•	•	•	•	•				•	•	•	•	•	•
2,3,4,6-Tetrachlorophenol	58-90-2									•																					
Tetrachlorvinphos	961-11-5									•																					
Tetraethyl pyrophosphate	107-49-3									•																					
Thionazine	297-97-2									•																					
Thiophenol (Benzenethiol)	108-98-5									•																					
Toluene	108-88-3														•	•	•	•	•	•	•	•	•	•	•		•		•			
Toluene diisocyanate	584-84-9									•																					
o-Toluidine	95-53-4									•																					
Toxaphene	8001-35-2				•	•	•	•	•	•	•	•	•																		
2,4,5-TP(Silvex)	93-72-1	•	•																												
1,2,3-Trichlorobenzene	87-61-6															•		•		•			•				•		•			
1,2,4-Trichlorobenzene	120-82-1									•		•	•	•		•		•		•			•				•		•			
2,4,5-Trichlorobiphenyl	15862-07-4									•																					
1,1,1-Trichloroethane	71-55-6														•		•	•	•	•	•	•					•	•	•	•	•	•
1,1,2-Trichloroethane	79-00-5														•		•	•	•	•	•	•					•	•	•	•	•	•
Trichloroethene	79-01-6														•		•	•	•	•	•	•				•	•	•	•	•	•
Trichlorofluoromethane	75-69-4														•		•	•		•							•	•	•	•	•	•
2,4,5-Trichlorophenol	95-95-4									•				•																	
2,4,6-Trichlorophenol	88-06-2									•		•	•	•																	
1,2,3-Trichloropropane	96-18-4														•		•	•		•							•	•	•	•		
0,0,0-Triethylphosphorothioate	126-68-1									•																					
Trifluralin	1582-09-8					•	•			•																					
Trimethyl phosphate	512-56-1									•																					
2,4,5-Trimethylaniline	137-17-7									•																					
1,2,4-Trimethylbenzene	95-63-6															•		•		•			•				•		•			
1,3,5-Trimethylbenzene	108-67-8															•		•		•			•				•		•			
1,3,5-Trinitrobenzene	99-35-4									•																					
Tri-p-tolyl phosphate(h)	78-32-0									•																					
Tris(2,3-dibromopropyl) phosphate	126-72-7									•																					
Vinyl acetate	108-05-4														•																	
Vinyl chloride	75-01-4														•		•	•	•	•	•	•					•	•	•	•	•	•
m-Xylene	108-38-3															•	•	•		•			•				•		•			
o-Xylene	95-47-6															•	•	•		•			•				•		•			
p-Xylene	106-42-3															•	•	•		•			•				•		•			
Xylene (total)	1330-20-7														•						•	•	•									

APPENDIX C

Inorganic analytes mentioned in the categories of Trace Metals identified by Flame and Graphite Furnace Atomic Absorption Spectroscopy, and Trace Metals identified by ICP are listed in alphabetical order. If the analyte is included in one of the listed test methods it is noted with a "•" in the appropriate column.

Determination of Total Trace Metals Cross-Reference

Metal	Flame and Graphite Furnace AA Spect.				ICP			
	SW-846 7000 Series	EPA 200 Series	Std Methods 3000 Series	CLP Inorganic Method	SW-846 6010A	EPA 200.7	Std. Methods 3120	CLP Inorganic Method
Aluminum	•	•	•	•	•	•	•	•
Antimony	•	•	•	•	•	•	•	•
Arsenic	•	•	•	•	•	•	•	•
Barium	•	•	•	•	•	•	•	•
Beryllium	•	•	•	•	•	•	•	•
Bismuth			•					
Boron						•	•	
Cadmium	•	•	•	•	•	•	•	•
Calcium	•	•	•	•	•	•	•	•
Cesium			•					
Chromium	•	•	•	•	•	•	•	•
Cobalt	•	•	•	•	•	•	•	•
Copper	•	•	•	•	•	•	•	•
Gold		•	•					
Iridium		•	•					
Iron	•	•	•	•	•	•	•	•
Lead	•	•	•	•	•	•	•	•
Lithium	•		•		•		•	
Magnesium	•	•	•	•	•	•	•	•
Manganese	•	•	•	•	•	•	•	•
Mercury[a]	•	•	•	•				
Molybdenum	•	•	•		•	•	•	
Nickel	•	•	•	•	•	•	•	•
Osmium	•	•	•					
Palladium		•	•					
Phosphorus					•			
Platinum		•	•					
Potassium	•	•	•	•	•	•	•	•
Rhenium		•	•					
Rhodium		•	•					
Ruthenium		•	•					
Selenium	•	•	•	•	•	•	•	•
Silicon / Silica (SiO$_2$)			•			•	•	
Silver	•	•	•	•	•	•	•	•
Sodium	•	•	•	•	•	•	•	•

Inorganic Analyte Cross-Reference to Methods

Determination of Total Trace Metals Cross-Reference (Continued)

Metal	Flame and Graphite Furnace AA Spect.				ICP			
	SW-846 7000 Series	EPA 200 Series	Std Methods 3000 Series	CLP Inorganic Method	SW-846 6010A	EPA 200.7	Std. Methods 3120	CLP Inorganic Method
Strontium	•		•		•		•	
Thallium	•	•	•	•	•	•	•	•
Thorium			•					
Tin	•	•	•					
Titanium		•	•					
Vanadium	•	•	•	•	•	•	•	•
Zinc	•	•	•	•	•	•	•	•

(a) Cold vapor technique.

APPENDIX D

Contained in this section are definitions and explanations of terms and abbreviations used in this book or commonly in use in this field.

AA. Atomic Absorption.

Absorb. To soak up. The incorporation of a liquid into a solid substance, as by capillary, osmotic, solvent, or chemical action. See Adsorb.

Absorbance. A measure of the decrease in incident light passing through a sample into the detector. It is defined mathematically as:

$$A = \frac{I \text{ (solvent)}}{I \text{ (solution)}} = \log \frac{I_o}{I}$$

Where, I = radiation intensity

Accuracy. Accuracy means the nearness of a result or the mean ($\bar{x}$) of a set of results to the true value. Accuracy is assessed by means of reference samples and percent recoveries.

Acid. An inorganic or organic compound that 1) reacts with metals to yield hydrogen; 2) reacts with a base to form a salt; 3) dissociates in water to yield hydrogen ions; 4) has a pH of less than 7.0; and 5) neutralizes bases or alkalies. All acids contain hydrogen and turn litmus paper red. They are corrosive to human tissue and are to be handled with care. See Base; pH.

Adsorb. To attract and retain gas or liquid molecules on the surface of another material. See Absorb.

Aliquot. A measured portion of a sample taken for analysis.

Alkali. Any compound having highly basic properties; i.e., one that readily ionizes in aqueous solution to yield OH anions, with a pH above 7, and turns litmus paper blue. Examples are oxides and hydroxides of certain metals belonging to group IA of the periodic table (Li, Na, K, Rb, Cs, Fr). Ammonia and amines may also be alkaline. Alkalies are caustic and dissolve human tissue. Treat alkali burns by quickly washing the affected area with large amounts of water for at least 15 min. Common commercial alkalies are sodium carbonate (soda ash), caustic soda and caustic potash, lime, lye, waterglass, regular mortar, Portland cement, and bicarbonate of soda. See Acid; Base; pH.

Ambient. Usual or surrounding conditions of temperatures, humidity, etc.

Analysis Date/Time. The date and military time (24 hour clock) of the introduction of the sample, standard, or blank into the analysis system.

Analyte. The element or ion compound an analyst seeks to determine; the element of interest.

Analytical Batch. The basic unit for analytical quality control is the analytical batch. The analytical batch is defined as samples which are analyzed together with the same method sequence and the same lots of reagents and with the manipulations common to each sample within the same time period or in continuous sequential time periods. Samples in each batch should be of similar composition (e.g. ground water, sludge, ash, etc.)

Analytical Sample. Any solution or media introduced into an instrument on which an analysis is performed excluding instrument calibration, initial calibration verification, initial calibration blank, continuing calibration verification and continuing calibration blank. Note the following are all defined as analytical samples: undiluted and diluted samples (EPA and non-EPA), predigestion spike samples, duplicate samples, serial dilution samples, analytical spike samples, post-digestion spike samples, interference check samples (ICS), CRDL standard for AA (CRA), CRDL standard for ICP (CRI), laboratory control sample (LCS), preparation blank (PB) and linear range analysis sample (LRS).

Analytical Spike. (Inorganic Analysis) The post-digestion spike. The addition of a known amount of standard after digestion.

ASTM. American Society for Testing and Materials. An organization that devises consensus standards for materials characterization and use. (1916 Race Street, Philadelphia, PA 19103; [215] 299-5400.)

atm. Atmosphere. A unit of pressure equal to the average pressure the air exerts at sea level. One atm = 1.013×10^5 N/m^2, or 14.7 lb/in.^{2}, or 760 mm Hg. Generally used in connection with high pressures.

Autozero. (Inorganic Analysis) Zeroing the instrument at the proper wavelength. It is equivalent to running a standard blank with the absorbance set at zero.

Average Intensity. (Inorganic Analysis) The average of two different injections (exposures).

Background Correction. (Inorganic Analysis) A technique to compensate for variable background contribution to the instrument signal in the determination of trace elements.

Bar Graph Spectrum. (Organic Analysis) A plot of the mass-to-charge ratio (m/e) versus relative intensity of the ion current.

Base. Substances that (usually) liberate OH anions when dissolved in water. Bases react with acids to form salts and water. Bases have a pH > 7, turn litmus paper blue, and may be corrosive to human tissue. A strong base is called alkaline or caustic. Examples are lye and DRANO™. See Acid; Alkali; pH.

Batch. A group of samples prepared at the same time in the same location using the same method.

Bias. Consistent deviation of measured values from the true value, caused by systematic errors in a procedure.

Blank. (also see Method Blank) A blank is an artificial sample designed to monitor the introduction of artifacts into the process. For aqueous samples, reagent water is used as a blank matrix; however, a universal blank matrix does not exist for solid samples, but sometimes clean sand is used as a blank matrix. The blank is taken through the appropriate steps of the process. A reagent blank is an aliquot of analyte-free water or solvent analyzed with the analytical batch. Field blanks are aliquots of analyte-free water or solvents brought to the field in sealed containers and transported back to the laboratory with the sample containers. Trip blanks and equipment blanks are two specific types of field blanks. Trip blanks are not opened in the field. They are a check on sample contamination originating from sample transport, shipping and from site conditions. Equipment blanks are opened in the field and the contents are poured appropriately over or through the sample collection device, collected in a sample container, and returned to the laboratory as a sample. Equipment blanks are a check on sampling device cleanliness.

B/N/A. Base, Neutral, Acid Extractable Compounds

Boiling Point, BP. The temperature at which a liquid's vapor pressure equals the surrounding atmospheric pressure so that the liquid rapidly vaporizes. Flammable materials with low BPs generally present special fire hazards [e.g., butane, BP = 31°F (-0.5°C); gasoline, BP = 100°F (38°C)]. For mixtures, a range of temperature is given.

BP. See Boiling Point.

4-Bromofluorobenzene (BFB). (Organic Analysis) Compound chosen to establish mass spectral instrument performance for volatile analyses. Also used as a surrogate for volatile organic analysis.

Buffer. A substance that reduces the change in hydrogen ion concentration (pH) otherwise produced by adding acids or bases to a solution. A pH stabilizer.

°C. Degrees Celsius (centigrade). Metric temperature scale on which 0 = water's freezing point and 100 = its boiling point. °F = (°C x 9/5) + 32 °C = (°F - 32) x 5/9. See °F.

Calibration. (Inorganic Analysis) The establishment of an analytical curve based on the absorbance, emission intensity, or other measured characteristic of known standards. The calibration standards must be prepared using the same type of acid and reagents or concentration of acids as used in the sample preparation.

Calibration Blank. (Inorganic Analysis) Usually an organic or aqueous solution that is as free of analyte as possible and prepared with the same volume of chemical reagents used in the preparation of the calibration standards and diluted to the appropriate volume with the same solvent (water or organic) used in the preparation of the calibration standard. The calibration blank is used to give the null reading for the instrument response versus concentration calibration curve. One calibration blank should be analyzed with each analytical batch or every method specified number of samples, whichever is greater.

Calibration Check. Verification of the ratio of instrument response to analyte amount, a calibration check is done by analyzing for analyte standards in an appropriate solvent. Calibration check solutions are made from a stock solution which is different from the stock used to prepare standards.

Calibration Check Standard. Standard used to determine the state of calibration of an instrument between periodic recalibrations.

Calibration Standards. A series of known standard solutions used by the analyst for calibration of the instrument (i.e., preparation of the analytical curve).

CAS Number (CAS Registration Number). An assigned number used to identify a chemical. CAS stands for Chemical Abstracts Service, an organization that indexes information published in *Chemical Abstracts* by the American Chemical Society and that provides index guides by which information about particular substances may be located in the abstracts. Sequentially assigned CAS numbers identify *specific* chemicals, except when followed by an asterisk (*) which signifies a compound (often naturally occurring) of variable composition. The numbers have no chemical significance. The CAS number is a concise, unique means of material identification. (Chemical Abstracts Service, Division of American Chemical Society, Box 3012, Columbus, OH 43210; [614] 447-3600.)

Caustic. See Alkali.

CCC. Calibration Check Compound.

CERCLA. The Comprehensive Environmental Response, Compensation, and Liability Act. The Superfund Law, Public Law PL 96-510, found at 40 CFR 300. The EPA has jurisdiction. Enacted Dec. 11, 1980, and amended thereafter, CERCLA provides for identification and cleanup of hazardous materials released over the land and into air, waterways, and groundwater. It covers areas affected by newly released materials and older leaking or abandoned dump sites. Report releases of hazardous materials to the National Response Center, (800) 424-8802. CERCLA established the superfund, a trust fund to help pay for cleanup of hazardous materials sites. The EPA has authority to collect cleanup costs from those who release the waste material. Cleanup funds come from fines and penalties, from taxes on chemical/petro-chemical feed stocks, and the US Treasury Dept. A separate fund collects taxes on active disposal sites to finance monitoring after they close. CERCLA is a result of the serious problems that arose from release of hazardous materials in Love Canal area near Niagara Falls, NY, in August 1978.

CFR. *Code of Federal Regulations*. A collection of the regulations established by law. Contact the agency that issued the regulation for details, interpretations, etc. Copies are sold by the Superintendent of Documents, Government Printing Office, Washington, DC 20402; (202) 783-3238.

Characterization. A determination of the approximate concentration range of compounds of interest used to choose the appropriate analytical protocol.

Check Sample. A blank which has been spiked with the analyte(s) from an independent source in order to monitor the execution of the analytical method is called a check sample. The level of the spike shall be at the regulatory action level when applicable. Otherwise, the spike shall be at 5 times the estimate of the quantification limit. The matrix used shall be phase matched with the samples and well characterized: for example, reagent grade water is appropriate for an aqueous sample.

Check Standard. A material of known composition that is analyzed concurrently with test samples to evaluate a measurement process. An analytical standard that is analyzed to verify the calibration of the analytical system. One check standard should be analyzed with each analytical batch or every 20 samples, whichever is greater.

(See Calibration Check Standard and Calibration Check).

CHEMTREC. Chemical Transportation Emergency Center. Established in Washington, DC, by the Chemical Manufacturers Association (CMA) to provide emergency information on materials involved in transportation accidents. Twenty-four-hour number: (800) 424-9300. In Washington, DC, Alaska, and Hawaii call (202) 483-7616.

CLP. Contract Laboratory Program.

Code of Federal Regulations. See CFR.

Coefficient of Variation (CV). The standard deviation as a percent of the arithmetic mean.

Combustible. A term the NFPA, DOT, and others use to classify certain materials with low flash points that ignite easily. Both NFPA and DOT generally define *combustible liquids* as having a flash point of 100°F (38°C) or higher. The NFPA classifies nonliquid materials such as wood and paper as *ordinary combustibles*. OSHA defines *combustible liquid* within the Hazard Communication Law as any liquid with a flash point at or above 100°F (38°C) but below 200°F (93.3°C). See Flammable.

Confidence Coefficient. The probability, %, that a measurement result will lie within the confidence interval or between the confidence limits.

Confidence Interval. Set of possible values within which the true value will lie with a specified level of probability.

Confidence Limit. One of the boundary values defining the confidence interval.

Corrosive. A chemical that causes visible destruction of or irreversible alterations in living tissue by chemical action at the site of contact, or which causes a severe corrosion rate in steel or aluminum. A waste that exhibits a "characteristic of corrosivity (40 CFR 261.22)," as defined by RCRA, may be regulated by EPA as a hazardous waste.

Continuing Calibration. (Organic Analysis) Analytical standard run every 12 hours to verify the calibration of the GC/MS system.

Continuing Calibration. (Inorganic Analysis) Analytical standard run every 10 analytical samples or every 2 hours, whichever is more frequent, to verify the calibration of the analytical system.

Continuous Liquid-Liquid Extraction. Used herein synonymously with the terms continuous extraction, continuous liquid extraction and liquid extraction. This extraction technique involves boiling the extraction solvent in a flask and condensing the solvent above the aqueous sample. The condensed solvent drips through the sample, extracting the compounds of interest from the aqueous phase.

Contract Required Detection Limit (CRDL). Minimum level of detection acceptable under the contract Statement of Work.

Control Limits. A range within which specified measurement results must fall to be compliant. Control limits may be mandatory, requiring corrective action if exceeded, or advisory, requiring that noncompliant data be flagged.

Correlation Coefficient. A number (r) which indicates the degree of dependence between two variables (concentration - absorbance). The more dependent they are the closer the value to one. Determined on the basis of the least squares line.

CRQL. Contract Required Quantitation Limit.

Cryogenic. Relating to extremely low temperature as for refrigerated gases.

cu ft, ft^3. Cubic foot. Cu ft is more usual.

cu m, m^3. Cubic meter. m^3 is preferred.

CVAA. Cold Vapor Atomic Absorption.

CWA. Clean Water Act. Public Law PL 92-500. Found at 40 CFR 100-140 and 400-470. Effective November 18, 1972, and amended significantly since then. EPA and Army Corps of Engineers have jurisdiction. CWA regulates the discharge of nontoxic and toxic pollutants into surface waters. Its ultimate goal is to eliminate all discharges into surface waters. Its interim goal is to make surface waters usable for fishing, swimming, etc. EPA sets guidelines, and states issue permits (NPDES, Natural Pollutant Discharge Elimination System permit) specifying the types of control equipment and discharges for each facility.

Decafluorotriphenylphosphine (DFTPP). (Organic Analysis) Compound chosen to establish mass spectral instrument performance for semivolatile analysis.

Density. Ratio of weight (mass) to volume of a material, usually in grams per cubic centimeter or pounds per gallon. See Specific Gravity.

Digestion Log. (Inorganic Analysis) An official record of the sample preparation (digestion).

Dissolved Metals. (Inorganic Analysis) Analyte elements which have not been digested prior to analysis and which will pass through a 0.45 μm filter.

DOT. US Dept. of Transportation. Regulates transportation of materials to protect the public as well as fire, law, and other emergency-response personnel. DOT classifications specify the use of appropriate warnings, such as Oxidizing Agent or Flammable Liquid. (400 7th St., SW, Washington, DC 20590.)

DOT Identification Numbers. Four-digit numbers used to identify particular materials for regulation of their transportation. See DOT publications that describe the regulations (49 CFR 172.102). These numbers are called product identification numbers (PINs) under the Canadian Transportation of Dangerous Goods Regulations. Those numbers used internationally may carry a "UN" prefix (e.g., UN 1170, ethyl alcohol), but those used only North America have an "NA" prefix (e.g., NA 9121, ferric sulphate).

Dry Weight. The weight of a sample based on percent solids. The weight after drying in an oven. See Percent Moisture.

Duplicate. A second aliquot of a sample that is treated the same as the original sample in order to determine the precision of the method.

Duplicate Samples. Duplicate samples are two separate samples taken from the same source (i.e., in separate containers and analyzed independently).

EDL. Estimated Detection Limit.

EICP. Extracted Ion Current Profile.

EMSL. Environmental Monitoring Systems Laboratory.

Environmental Sample. An environmental sample or field sample is a representative sample of any material (aqueous, nonaqueous, or multimedia) collected from any source for which determination of composition or contamination is requested or required. Environmental samples are normally classified as follows:

Drinking Water -- delivered (treated or untreated) water designated as potable water;

Water/Wastewater -- raw source waters for public drinking water supplies, ground waters, municipal influents/effluents, and industrial influents/effluents;

Sludge -- municipal sludges and industrial sludges;

Waste -- aqueous and nonaqueous liquid wastes, chemical solids, contaminated soils, and industrial liquid and solid wastes.

EPA, (Canada) Environmental Protection Act. Federal legislation, administered by Environment Canada, designed to protect the environment.

EPA, (US) Environmental Protection Agency. A Federal agency with environmental protection regulatory and enforcement authority. Administers the CAA, CWA, RCRA, TSCA, and other Federal environmental laws. (400 M Street, SW, Washington, DC 20460; [202] 382-2090.)

Equipment Blank. Usually an organic or aqueous solution that is as free of analyte as possible and is transported to the site, opened in the field, and poured over or through the sample collection device, collected in a sample container, and returned to the laboratory. This serves as a check on sampling device cleanliness. One equipment blank should be analyzed with each analytical batch or every 20 samples, whichever is greater.

EQL Estimated Quantitation Limit.

Extractable. (Organic Analysis) A compound that can be partitioned into an organic solvent from the sample matrix and is amenable to gas chromatography. Extractables include semivolatile (BNA) and pesticide/Aroclor compounds.

°F *or* **F.** Degrees Fahrenheit. See °C.

Federal Register (US). See FR.

Field Blank. Usually an organic or aqueous solution that is as free of analyte as possible and is transferred from one vessel to another at the sampling site and preserved with the appropriate reagents. This serves as a check on reagent and environmental contamination. One field blank should be analyzed with each analytical batch or every 20 samples, whichever is greater.

Fire Diamond (NFPA Hazard Rating). Per "NFPA 704" publication. Visual system that provides a general idea of the inherent hazards, and their severity, of materials relating to *fire* prevention, exposure and control. Preferred reading order; Health, Flammability, Reactivity, Special.

Position A - Health Hazard (Blue). Degree of hazard; level of short-term protection

0 = Ordinary Combustible Hazards in a Fire

1 = Slightly Hazardous

2 = Hazardous

3 = Extreme Danger

4 = Deadly

Position B - Flammability (Red). Susceptibility to burning

0 = Will Not Burn

1 = Will Ignite if Preheated

2 = Will Ignite if Moderately Heated

3 = Will Ignite at Most Ambient Conditions

4 = Burns Readily at Ambient Conditions

Position C - Reactivity, Instability (Yellow). Energy released if burned, decomposed, or mixed

0 = Stable and Not Reactive with Water

1 = Unstable if Heated

2 = Violent Chemical Change

3 = Shock and Heat May Detonate

4 = May Detonate

Position D - Special Hazard (White).

OX = Oxidizer

W = Use No Water, reacts!

Fire Point. The lowest temperature at which a liquid produces sufficient vapor to flash near its surface and continues to burn. Usually 10 to 30°C higher than the flash point.

Flame Atomic Absorption (FLAA). Atomic absorption which utilizes flame for excitation.

Flammable. Describes any solid, liquid, vapor, or gas that ignites easily and burns rapidly. See Combustible and Inflammable.

Flammable Liquid. A liquid that gives off vapors readily ignitable at room temperature. Defined by the NFPA and DOT as a liquid with a flash point below 100°F (38°C).

Flash Point, FP. Lowest temperature at which a flammable liquid gives off sufficient vapor to form an ignitable mixture with air near its surface or within a vessel. Combustion does not continue. FP is determined by laboratory tests in cups. See Fire Point.

FR. The *Federal Register*. A daily publication that lists and discusses Federal regulations. Available from the Government Printing Office.

10% Frequency. A frequency specification during an analytical sequence allowing for no more than 10 analytical samples between required calibration verification measurements, as specified by the Contract Statement of Work.

FSCC. Fused Silica Capillary Column

g. Gram. Metric unit of weight. See kg.

GC/EC. Gas Chromatography/Electron Capture.

GC/MS. Gas Chromatography/Mass Spectrometry.

Graphite Furnace Atomic Absorption (GFAA). Atomic absorption which utilizes a graphite cell for excitation.

Holding Time. The elapsed time expressed in days from the date of receipt of the sample by the Contractor until the date of its analysis.

HRGC. High Resolution Gas Chromatography

HRMS. High Resolution Mass Spectrometry

ICP/MS. Inductively Coupled Plasma/Mass Spectrometry

ICS. Interference Check Standard.

ID. Identification.

IDL. Instrument Detection Limit.

Independent Standard. A contractor-prepared standard solution that is composed of analytes from a different source than those used in the standards for the initial calibration.

Inductively Coupled Plasma (ICP). A technique for the simultaneous or sequential multi-element determination of elements in solution. The basis of the method is the measurement of atomic and ionic emission by an optical spectroscopic technique. Characteristic atomic and ionic line emission spectra are produced by excitation of the sample in a radio frequency inductively coupled plasma.

In-House. At the Contractor's facility.

Initial Calibration. Analysis of analytical standards for a series of different specified concentrations; used to define the linearity and dynamic range of the response of the analytical detector or method.

Injection. Introduction of the analytical sample into the instrument excitation system for the purpose of measuring absorbance, emission or concentration of analyte. May also be referred to as exposure.

Instrument Calibration. Analysis of analytical standards for a series of different specified concentrations; used to define the quantitative response, linearity, and dynamic range of the instrument to target analytes.

Instrument Check Sample. (Inorganic Analysis) A solution containing both interfering and analyte elements of known concentration that can be used to verify background and interelement correction factors.

Instrument Check Standard. A multi-element standard of known concentration prepared by the analyst to monitor and verify instrument performance on a daily basis.

Instrument Detection Limit (IDL). Determined by multiplying by three the standard deviation obtained for the analysis of a standard solution (each analyte in reagent water) at a concentration of 3 times to 5 times IDL on three nonconsecutive days with seven consecutive measurements per day.

Interferents. Substances which affect the analysis for the element of interest.

Internal Standard(s). Compound(s) added to every standard, blank, matrix spike, matrix spike duplicate, sample (for VOAs), and sample extract (for semivolatiles) at a known concentration, prior to analysis. Internal standard(s) are used as the basis for quantitation of the target compounds.

IS. Internal Standard.

Isomers. Chemical compounds with the same molecular weight and atomic composition but differing molecular structure; e.g., *n*-pentane and 2-methylbutane.

kg, kilogram. 1000 grams.

l. Liter. Basic metric unit of volume. One liter of water weighs 1 kg and is equal to 1.057 quarts.

Laboratory Control Sample (LCS). A control sample of known composition. Aqueous and solid laboratory control samples are analyzed using the same sample preparation, reagents, and analytical methods employed for the EPA samples received.

Laboratory Control Standard. A standard, usually certified by an outside agency, used to measure the bias in a procedure. For certain constituents and matrices, use National Institute of Standards and Technology (NIST) Standard Reference Materials when they are available.

Limit of Quantitation (LOQ). The constituent concentration that produces a signal sufficiently greater than the blank that it can be detected with the specified limits by good laboratories during routine operating conditions. Typically it is the concentration that produces a signal 10σ above the reagent water blank signal.

Linear Range, Linear Dynamic Range. (Inorganic Analysis) The concentration range over which the ICP analytical curve remains linear.

Lower Limit of Detection (LLD). The constituent concentration in reagent water that produces a signal $2(1.645)\sigma$ above the mean of blank analyses. This sets both Type I and Type II errors at 5%. Other names for this limit are "detection limit" and "limit of detection (LOD)".

LSE. Liquid-Solid Extraction.

m. Meter. Metric unit of length equal to 39.37 in.

m^3 or cu m. Cubic meter; m^3 is preferred.

Material Safety Data Sheet. See MSDS.

Matrix. The predominant material of which the sample to be analyzed is composed. Matrix is not synonymous with phase (liquid or solid).

Matrix Modifier. (Inorganic Analysis) Reagents used in AA to lessen the effects of chemical interferents, viscosity, and surface tension.

Matrix Spike (MS). Aliquot of a sample (water or soil) fortified (spiked) with known quantities of specific compounds and subjected to the entire analytical procedure in order to indicate the appropriateness of the method for the matrix by measuring recovery.

Matrix Spike Duplicate (MSD). A second aliquot of the same matrix as the matrix spike (above) that is spiked in order to determine the precision of the method.

MDL. Method Detection Limit

Metabolism. The chemical and physical processes whereby the body functions.

Meter (m). The basic metric measure of length; equivalent to 39.371 in.

Method Blank. An analytical control consisting of all reagents, internal standards and surrogate standards, that is carried through the entire analytical procedure. The method blank is used to define the level of laboratory background and reagent contamination.

Method Detection Limit (MDL). The constituent concentration that, when processed through the complete method, produces a signal with a 99% probability that it is different from the blank. For seven replicates of the sample, the mean must be 3.14 above the blank where σ is the standard deviation of the seven replicates. The MDL will be larger than the LLD because of the few replications and the sample processing steps and may vary with constituent and matrix.

Method of Standard Additions (MSA). (Inorganic Analysis) The addition of 3 increments of a standard solution (spikes) to sample aliquots of the same size. Measurements are made on the original and after each addition. The slope, x-intercept and y-intercept are determined by least-square analysis. The analyte concentration is determined by the absolute value of the x-intercept. Ideally, the spike volume is low relative to the sample volume (approximately 10% of the volume). Standard addition may counteract matrix effects; it will not counteract spectral effects. Also referred to as Standard Addition.

mg. Milligram (1/1000, 10^{-3}, of a gram).

Microgram (μg). One-millionth (10^{-6}) of a gram.

Micrometer (μm). One-millionth (10^{-6}) of a meter; often referred to as a micron.

Millimeter (mm). 1/1000 of a meter.

min. Minute

Miscible. When two liquids or two gases are completely soluble in each other in all proportions. While gases mix with one another in all proportions, the miscibility of liquids depends on the chemical natures.

ml. Milliliter. One thousandth of a liter. A metric unit of capacity, for all practical purposes equal to 1 cubic centimeter. One cubic inch is about 16 ml.

Mole *or* **mol.** The quantity of a chemical substance that has a mass in grams numerically equal to the molecular weight. For example, salt (NaCl) has a molecular weight of 58.5 (Na, 23, and chlorine, 35.5). Thus, one mole of NaCl is 58.5 g.

Molecular Weight. The sum of atomic weights of the atoms in a molecule. For example, water (H_2O) has a molecular weight of 18.015, the atomic weights being hydrogen = 2 (1.008) + oxygen = 15.999.

MQL. The method quantification limit (MQL) is the minimum concentration of a substance that can be measured and reported.

MSDS. Material safety data sheet. OSHA has established guidelines for the descriptive data that should be concisely provided on a data sheet to serve as the basis for written hazard communication programs. The thrust of the law is to have those who make, distribute, and use hazardous materials responsible for effective communication. See the Hazard Communication Rule, 29 CFR, Part 1910. 1200, as amended, Sec. g. See Schedule I, Sec. 12, of the Canadian Hazardous Products Act.

MW. See Molecular Weight.

m/z. (Organic Analysis) Mass to charge ratio, synonymous with "m/e".

n-. Normal. A chemical name prefix signifying a straight-chain structure; i.e., no branches.

NA, ND. Not applicable, not available; not determined.

Narrative. A descriptive documentation of any problems encountered in processing the samples, along with corrective action taken and problem resolution.

ng. Nanogram. One billionth, 10^{-9}, of a gram.

NIOSH. National Institute of Occupational Safety and Health. The agency of the Public Health Service that tests and certifies respiratory and air-sampling devices. It recommends exposure limits to OSHA for substances, investigates incidents, and researches occupational safety. (NIOSH, 4676 Columbia Parkway, Cincinnati, OH 45226; [513] 533-8328.)

NIST. National Institute for Standards and Technology

NOC. Not otherwise classified.

Occupational Safety and Health Act. See OSH Act.

Occupational Safety and Health Administration. See OSHA.

OSHA. The Occupational Safety and Health Administration. Part of the US Department of Labor. The regulatory and enforcement agency for safety and health in most US industrial sectors. (Documents are available from the OSHA Technical Data Center Docket Office, Rm N-3670, 200 Constitution Ave, NW, Washington, DC 20210; [202] 523-7894.)

OSH Act. The Occupational Safety and Health Act of 1970. Effective April 28, 1971. Public Law 91-596. Found at 29 CFR 1910, 1915, 1918, 1926. OSHA jurisdiction. The regulatory vehicle to ensure the safety and health of workers in firms larger than 10 employees. Its goal is to set standards of safety that prevent injury and illness among the workers. Regulating employee exposure and informing employees of the dangers of materials are key factors. This act established the Hazard Communication Rule (29 CFR 1910.1200). See Hazard Communication Rule for details.

PE. Performance Evaluation

Percent Moisture. An approximation of the amount of water in a soil/sediment sample made by drying an aliquot of the sample at 105°C. The percent moisture determined in this manner also includes contributions from all compounds that may volatilize at or below 105°C, including water. Percent moisture may be determined from decanted samples and from samples that are not decanted.

Percent Solids. The proportion of solid in a soil sample determined by drying an aliquot of the sample.

Performance Evaluation (PE) Sample. A sample of known composition provided by EPA for Contractor analysis. Used by EPA to evaluate Contractor performance.

PEST. Pesticides

pH. "Hydrogen ion exponent," a measure of hydrogen ion concentration. A scale (0 to 14) representing an aqueous solution's acidity or alkalinity. Low pH values indicate acidity and high values, alkalinity. The scale's mid-point, 7, is neutral. Substances in an aqueous solution ionize to various extents giving different concentrations of H and OH ions. Strong acids have excess H ions and a pH of 1 to 3 (HCl, pH = 1). Strong bases have excess OH ions and a pH of 11 to 13 (NaOH, pH = 12).

ppb. Parts per billion.

ppm. Parts per million. "Parts of vapor or gas per million parts of contaminated air by volume at 25°C and 1 torr pressure" (ACGIH). At 25°C, ppm = $(mg/m^3 \times 24.45)$ divided by molecular weight.

ppt. Parts per trillion.

PQL. The practical quantitation limit (PQL) is the lowest level that can be reliably achieved within specified limits of precision and accuracy during routine laboratory operating conditions.

Precision. Precision is the agreement between a set of replicate measurements without assumption of knowledge of the true value. Precision is assessed by means of duplicate/replicate sample analysis.

Preparation Blank (reagent blank, method blank). An analytical control that contains distilled, deionized water and reagents, which is carried through the entire analytical procedure (digested and analyzed). An aqueous method blank is treated with the same reagents as a sample with a water matrix; A solid method blank is treated with the same reagents as a soil sample.

Protocol. Describes the exact procedures to be followed with respect to sample receipt and handling, analytical methods, data reporting and deliverables, and document control. Used synonymously with Statement of Work (SOW).

Purge and Trap (Device). Analytical technique (device) used to isolate volatile (purgeable) organics by stripping the compounds from water or soil by a stream of inert gas, trapping the compounds on an adsorbent such as a porous polymer trap, and thermally desorbing the trapped compounds onto the gas chromatographic column.

QA. Quality Assurance

QAP. Quality Assurance Plan

QC. Quality Control

Quality Assessment. Procedure for determining the quality of laboratory measurements by use of data from internal and external quality control measures.

Quality Assurance. A definitive plan for laboratory operation that specifies the measures used to produce data of known precision and bias.

Quality Control. Set of measures within a sample analysis methodology to assure that the process is in control.

Quality Control Reference Sample. A sample prepared from an independent standard at a concentration other than that used for calibration, but within the calibration range. An independent standard is defined as a standard composed of the analyte(s) of interest from a different source than that used in the preparation of standards for use in the standard curve. A quality control reference sample is intended as an independent check of technique, methodology, and standards and should be run with every analytical batch or every 20 samples, whichever is greater. This is applicable to all organic and inorganic analyses.

Quality Control Sample. A solution obtained from an outside source having known concentration values to be used to verify the calibration standards.

R. Recovery.

Random Error. The deviation in any step in an analytical procedure that can be explained by standard statistical techniques.

RCRA. Resource Conservation and Recovery Act, PL 94-580. Found at 40 CFR 240-271. EPA has jurisdiction. Enacted November 21, 1976, and amended since. RCRA's major emphasis is the control of hazardous waste disposal. It controls all solid-waste disposal and encourages recycling and alternative energy sources. In 1984 the USA generated 265 million tons of hazardous waste.

RCRA Hazardous Waste. A material designated by RCRA as hazardous waste and assigned a number to be used in record keeping and reporting compliance (e.g., D003, F001, U169).

Reagent Blank. Usually an organic or aqueous solution that is as free of analyte as possible and contains all the reagents in the same volume as used in the processing of the samples. The reagent blank must be carried through the complete sample preparation procedure and contains the same reagent concentrations in the final solution as in the sample solution used for analysis. The reagent blank is used to correct for possible contamination resulting from the preparation or processing of the sample. One reagent blank should be prepared for every analytical batch or for every 20 samples, whichever is greater.

Reagent Grade. Analytical reagent (AR) grade, ACS reagent grade, and reagent grade are synonymous terms for reagents which conform to the current specifications of the Committee on Analytical Reagents of the American Chemical Society.

Reagent Water. Water in which an interferent is not observed at or above the minimum quantitation limit of the parameters of interest.

Reconstructed Ion Chromatogram (RIC). (Organic Analysis) A mass spectral graphical representation of the separation achieved by a gas chromatograph; a plot of total ion current versus retention time.

Relative Percent Difference (RPD). To compare two values, the relative percent difference is based on the mean of the two values, and is reported as an absolute value, i.e., always expressed as a positive number or zero.

Relative Response Factor (RRF). A measure of the relative mass spectral response of an analyte compared to its internal standard. Relative Response Factors are determined by analysis of standards and are used in the calculation of concentrations of analytes in samples.

Replicate. Repeated operation occurring within an analytical procedure. Two or more analyses for the same constituent in an extract of a single sample constitutes replicate extract analyses.

Replicate Sample. A replicate sample is a sample prepared by dividing a sample into two or more separate aliquots. Duplicate samples are considered to be two replicates. In cases where aliquoting is impossible, as in the case of volatiles, duplicate samples must be taken for the replicate analysis.

Resolution. Also termed separation or percent resolution, the separation between peaks on a chromatogram, calculated by dividing the depth of the valley between the peaks by the peak height of the smaller peak being resolved, multiplied by 100.

Resource Conservation and Recovery Act. See RCRA.

Retention Time Window. Usually defined as three times the standard deviation of the absolute or relative RT of an analyte standard injected over the course of a 72 hour period.

Rounding Rules. If the figure following those to be retained is less than 5, the figure is dropped, and the retained figures are kept unchanged. As an example, 11.443 is rounded off to 11.44.

If the figure following those to be retained is greater than 5, the figure is dropped, and the last retained figure is raised by 1. As an example, 11.446 is rounded off to 11.45.

If the figure following those to be retained is 5, and if there are no figures other than zeros beyond the five, the figure 5 is dropped, and the last-place figure retained is increased by one if it is an odd number or it is kept unchanged if an even number. As an example, 11.435 is rounded off to 11.44, while 11.425 is rounded off to 11.42.

If a series of multiple operations is to be performed (add, subtract, divide, multiply), all figures are carried through the calculations. Then the final answer is rounded to the proper number of significant figures.

RRT. Relative Retention Time.

RSD. Relative Standard Deviation.

RT. Retention Time.

RT Window. See Retention Time Window.

Run. A continuous analytical sequence consisting of prepared samples and all associated quality assurance measurements as required by the Contract Statement of Work.

Sample. A portion of material to be analyzed that is contained in single or multiple containers and identified by a unique sample number.

Sample Delivery Group (SDG). Unit within a single case that is used to identify a group of samples for delivery. An SDG is a group of 20 or fewer field samples within a case, received over a period of up to 14 calendar days (7 calendar days for 14-day data turnaround contracts). Data from all samples in an SDG are due concurrently.

Sample Number (EPA Sample Number). A unique identification number designated by EPA for each sample. The EPA sample number appears on the sample Traffic Report which documents information on that sample.

SARA. Superfund Amendments and Reauthorization Act. Signed into law October 17, 1986. Title III of SARA is known as the Emergency Planning and Community Right-to-Know Act of 1986. A revision and extension of CERCLA, SARA is intended to encourage and support local and state emergency planning efforts. It provides citizens and local governments with information about potential chemical hazards in their communities. SARA calls for facilities that store hazardous materials to provide officials and citizens with data on the types (flammables, corrosives, etc.); amounts on hand (daily, yearly); and their specific locations. Facilities are to prepare and submit inventory lists, MSDSs, and tier 1 and 2 inventory forms. The disaster in Bhopal, India, in 1987 added impetus to the passage of this law.

SD. Standard Deviation.

Semivolatile Compounds. (Organic Analysis) Compounds amenable to analysis by extraction of the sample with an organic solvent. Used synonymously with Base/Neutral/Acid (BNA) compounds.

Sensitivity. The slope of the analytical curve, i.e., functional relationship between emission intensity and concentration.

Serial Dilution. The dilution of a sample by a factor of five. When corrected by the dilution fac-

tor, the diluted sample must agree with the original undiluted sample within specified limits. Serial dilution may reflect the influence of interferents.

SIM. Selected Ion Monitoring.

SOP. Standard Operation Procedure.

SOW. Statement of Work.

SPCC. System Performance Check Compound.

Specific Gravity. The ratio of the mass of a substance to the same volume of a reference substance, at a specified temperature. Specific gravity is a dimensionless number. Water (density 1 kg/l at 4°C) is the reference for solids and liquids, while air (density 1.29 g/l at 0°C and 760 mm Hg pressure) is the reference for gases. If a volume of a material weighs 8 g, and an equal volume of water weighs 10 g, the material has a specific gravity of 0.8 (8 ÷ 10 = .8). Insoluble materials with specific gravity greater than 1.0 sink to the bottom in water. Specific gravity is an important fire session and spill cleanup consideration since most (but not all) flammable liquids have a specific gravity less than 1.0 and, if insoluble, float on water.

Standard Analysis. An analytical determination made with known quantities of target compounds; used to determine response factors.

Standard Curve. A standard curve is a curve which plots concentrations of known analyte standards versus the instrument response to the analyte. Calibration standards are prepared by diluting the stock analyte solution in graduated amounts which cover the expected range of the samples being analyzed. Standards should be prepared at the frequency specified in the appropriate section. The calibration standards must be prepared using the same type of acid or solvent and at the same concentration as will result in the samples following sample preparation. This is applicable to organic and inorganic chemical analyses. See Calibration.

Stock Solution. Standard solution which can be diluted to derive other standards.

Superfund Amendments and Reauthorization Act. See SARA, CERCLA.

Surrogate. Surrogates are organic compounds which are similar to analytes of interest in chemical composition, extraction, and chromatography, but which are not normally found in environmental samples. These compounds are spiked into all blanks, calibration and check standards, samples (including duplicates and QC reference samples) and spiked samples prior to analysis. Percent recoveries are calculated for each surrogate.

Surrogate Standard. A pure compound added to a sample in the laboratory just before processing so that the overall efficiency of a method can be determined.

Suspended. (Inorganic Analysis) Those elements which are retained by a 0.45 µm membrane filter.

SV. Semivolatile.

SVOA. Semivolatile Organic Analysis.

System Monitoring Compounds. (Organic Analysis) Compounds added to every blank, sample, matrix spike, matrix spike duplicate, and standard for volatile analysis, and used to evaluate the performance of the entire purge and trap-gas chromatograph-mass spectrometer system. These compounds are brominated or deuterated compounds not expected to be detected in environmental media.

Target Compound List (TCL). A list of compounds designated by the Statement of Work for analysis.

TCL. Target Compound List

Tentatively Identified Compounds (TIC). Compounds detected in samples that are not target compounds, internal standards, system monitoring compounds, or surrogates. Up to 30 peaks (those greater than 10% of peak areas or heights of nearest internal standards) are subjected to mass spectral library searches for tentative identification.

TIC. Tentatively Identified Compound

Time. When required to record time on any deliverable item, time shall be expressed as Military Time, i.e., a 24-hour clock.

Total Metals. (Inorganic Analysis) Analyte elements which have been digested prior to analysis.

Toxic Substances Control Act. See TSCA.

Traffic Report (TR). An EPA sample identification form filled out by the sampler, which accompanies the sample during shipment to the laboratory and which is used for documenting sample condition and receipt by the laboratory.

Trip Blank. Usually an organic or aqueous solution that is as free of analyte as possible and is transported to the sampling site and returned to the laboratory without being opened. This serves as a check on sample contamination originating from sample transport, shipping, and from the site conditions. One trip blank should be analyzed with each analytical batch or every 20 samples, whichever is greater.

TSCA. Toxic Substances Control Act. Public Law PL 94-469. Found in 40 CFR 700-799. EPA has jurisdiction. Effective Jan. 1, 1977. Controls the exposure to and use of raw industrial chemicals not subject to other laws. Chemicals are to be evaluated prior to use and can be controlled based on risk. The act provides for a listing of all chemicals that are to be evaluated prior to manufacture or use in the US. (Call the EPA, Industry Assistance Office, [202] 554-1404.)

Twelve-Hour Time Period. (Organic Analysis) The twelve (12) hour time period for GC/MS system instrument performance check, standards calibration (initial or continuing calibration), and method blank analysis begins at the moment of injection of the DFTPP or BFB analysis that the laboratory submits as documentation of instrument performance. The time period ends after 12 hours have elapsed according to the system clock. For pesticide/Aroclor analysis performed by GC/EC, the twelve hour time period in the analytical sequence begins at the moment of injection of the instrument blank that precedes sample analyses, and ends after twelve hours have elapsed according to the system clock.

Type I Error. Also called alpha error, is the probability of deciding a constituent is present when it actually is absent.

Type II Error. Also called beta error, is the probability of not detecting a constituent when it actually is present.

Validated Time of Sample Receipt (VTSR). The date on which a sample is received at the Contractor's facility, as recorded on the shipper's delivery receipt and Sample Traffic Report.

VOA. Volatile Organics Analysis.

Volatile Compounds. (Organic Analysis) Compounds amenable to analysis by the purge and trap technique. Used synonymously with purgeable compounds.

Water. Reagent, analyte-free, or laboratory pure water means distilled or deionized water or Type II reagent water which is free of contaminants that may interfere with the analytical test in question.

Wet Weight. The weight of a sample aliquot including moisture (undried).

Wide Bore Capillary Column. A gas chromatographic column with an internal diameter (ID) that is greater than 0.32 mm. Columns with lesser diameters are classified as narrow bore capillaries.

APPENDIX E

Should you need assistance interpreting a regulation or understanding the details of a method, you can get help from the Federal and state agencies listed in this appendix.

NATIONAL AND REGIONAL CONTACTS

ENVIRONMENTAL HOTLINES

Asbestos Ombudsman Clearinghouse/ Hotline
All except VA.............................. (800) 368-5888
VA only .... (703) 557-1938
....FTS: 557-1938
TDD machine .... (703) 557-2824
Provides information to the public sector, including individual citizens and community services on the handling and abatement of asbestos in schools, the workplace and the home.

Chemical Manufacturers Association Hotlines
National...................................... (202) 887-1216
Non-emergency information on chemicals
Chemical Referral Center (CRC)
National...................................... (800) 262-8200
District of Columbia and Alaska...... (202) 887-1315
Provides Chemical Emergency Information.

CHEMTREC
National..................................... (800) 424-9300
District of Columbia...................... (202) 887-4620

Consumer Product Safety Commission Hotline
National..................................... (800) 638-2772

Emergency Planning and Community Right-to-Know (Title III SARA) Hotline
National .... (800) 535-0202
Virginia .... (703) 920-9877
Provides regulatory, policy and technical assistance to federal agencies, local and state governments, the public and regulated community in response to questions related to the Emergency Planning and Community Right-to-Know Act (Title III of SARA). Information on reporting of hazardous substances for community planning purposes.

EPA Hotline.............................800-621-8431

EPA Superfund (Region II Hotline)
Restricted area codes as follows: 809, 201, 609, 908, 906, 212, 315, 516, 518, 607, 716, 718, 914
.... (800) 245-2738
Enables the Superfund Civil Investigators to receive information relevant to specific Superfund Site Enforcement Investigations.

EPA Technical Resources and Federal Superfund Hotline
National................................... (800) 732 - 1223
Provides information on technical questions

about the federal Superfund program.

EPA Toxic Substance Control Act (TSCA) Hotline
National................................... (800) 424-9065
.... (202) 554-1404
Information on regulations, precautions and health effects of toxic substances.

Mercury Hotline
National................................... (800) 833-3505
Provides answers to questions. No emergency Service

National Pesticides Telecommunications Network (Pesticides and Herbicides Hotline)
National, incl. AK, PR, VI .... (800) 858-PEST
.... (800) 858-7378
.... (806) 743-3091
Provides the medical, veterinary, professional communities and general public with information on: pesticides and herbicides product information, recognition and management of pesticide poisonings, toxicology and symptomatic reviews, safety information, health and environmental effects, clean-up and disposal procedures.

National Radon Hotline
National................................... (800) 767-7230
.... (800) SOS-RADON
Radon testing information. A message records names and addresses of callers and a brochure on radon is sent via 1st class mail.

National Response Center - U.S. Coast Guard Oil and Hazardous Material Spills
National except DC .... (800) 424-8802
DC and outside U.S. .... (202) 426-2675
DC and outside U.S. .... (202) 267-2185
For reporting of oil and hazardous material spills.
NOTE: Please have ready as much relevant data as possible when calling.

Occupational Safety and Health Administration Hotline
National......................................(800) 321-6742
24-hour access line to report unsafe and hazardous work practices.

Office of the Inspector General - Whistle Blower Hotline
National .... (800) 424-4000
.... (202) 382-4977
Fraud, waste or mismanagement in EPA-funded activities.

RCRA/Superfund/Community Right-to-Know (Title III) Hotline
International .... (800) 424-9346
.... (703) 920-9810

TDD machine .... (800) 553-7672
Answers factual questions from the regulated
community, other interested parties and the pub-
lic about EPA's RCRA regulations and policies;
referrals for obtaining related documents. RCRA,
Underground Storage Tanks (USTs), Superfund/
CERCLA and Pollution Prevention/Waste
Minimization.

Risk Communication Hotline
National .... (202) 382-5606
....FTS: 382-5606
Responds to questions from EPA program offices
and regions, and external inquiries as time per-
mits, regarding the EPA's Risk Communication
Program and risk communication issues.

Safe Drinking Water Hotline
National .... (800) 426-4791
Provides assistance and regulatory knowledge to
the regulated community (public water systems)
and the public on the regulations and programs
developed in response to the Safe Drinking Water
Act Amendments of 1986.

Substance Identification Hotline
National .... (800) 848-6538
Identifies chemical by CAS number or Name

US Army Corps of Engineers
National .... (202) 272-0001

US Department of Transportation (U.S. DOT)
National .... (202) 366-4488
Information of DOT CFR-49 regulations.

OTHER REGIONAL AND NATIONAL CONTACTS

Centers for Disease Control (CDC)
National .... (404) 639-3535
Answers technical and public questions.

Environmental Business Council of the United States
National .... (617) 890-4242

Environmental Technologies Export Council (ETECH)
National .... (619) 456-1861

EPA Action Line
IA, MO, NE, KA residents only; ...... (800) 223-0425
Referral service to appropriate program office.

EPA Assistant Administrator for Enforcement and Compliance Monitoring
National .... (202) 260-4134

Provides direction for the review and enforcement of compliance activities.

EPA Assistant Administrator for Research and Development
National .... (202) 260-7676
Provides technical information for the EPA
administrator on scientific and technical issues.

EPA General Information
National .... (617) 565-3421
Referral service to appropriate program office.

EPA General Information - Environmental Issues
CO, MO, ND, SD, UT, WY
residents .... (800) 759-4372
.... (303) 293-1603
General information - environmental issues.

EPA Method Information Communication Exchange (MICE)
National .... (703) 821-4789
Provides information using a voice-mail answer-
ing system about technical questions regarding
testing methods contained in Test Methods for
Evaluating Solid Waste Physical/Chemical
Methods (SW-846).

EPA Public Information Service
National .... (202) 260-7751
Provides non-technical documents and guidance
about general environmental information for the
public.

Center for Hazardous Materials (CHMR)
National .... (800) 334-2467
.... (412) 826-5320
Regulatory, toxic waste minimization, pollution
prevention, publications and referrals.

Center for International Environmental Law
National ....(202) 332-4840

Control Technology Center (CTC)
National .... (919) 541-0800
....FTS: 629-0800
Technical support and guidance on air pollution
emissions and control technology. Air emissions
and air pollution control technology for all air
pollutants including air toxins emitted by stationary
sources.

Food and Drug Administration (FDA)
National .... (301) 443-1544

Hazardous Waste Ombudsman
National .... (800) 262-7937
Washington, DC .... (202) 475-9361
The hazardous waste management program
established under RCRA is a highly complex

regulatory program developed by EPA. It assists the public and regulated community in resolving problems concerning any program or requirement under the Hazardous Waste Program. The ombudsman handles complaints from citizens and the regulated community, obtains facts, sorts information, and substantiates policy.

Mobile Sources
New England states only .... (800) 821-1237
MA only .... (800) 631-2700
Complaints regarding auto emission tampering, emission, auto warranty, recall notices, fuel issues, CSC recycling, auto air conditioning.

National Institute for Occupational Safety and Health (NIOSH)
National....................................(800) 356-4674

National Institute of Standards and Technology (NIST)
National....................................(301) 975-2000
....(303) 497-3000
Previously the National Bureau of Standards (NBS).

National Technical Information Service (NTIS)
National....................................(703) 487-4600
Source of information services (technical documents and databases) and documents from federal agencies, industries, and universities.

Nuclear Regulatory Commission
National....................................(301) 492-7000
NRC provides information about technical questions and documents regarding hazardous materials and wastes.

Public Health Service
National....................................(301) 443-2403

Water and Waste Water Information (600-series methods)
National....................................(202) 260-7120
Provides information on testing methods for water and waste water (600-series methods) contained in CFR-40 part 136.

Wetlands Protection
National, VI and Guam.................. (800) 832-7828
Responsive to public interest, questions and requests for information about the values and functions of wetlands and options for their protection. Provides referrals to callers when necessary.

ENVIRONMENTAL DATABASE AND COMPUTER CONTACTS

CERCLIS - Helpline
National .... (202) 353-0056
....FTS: 252-0056
Answering machine for all off-hour callers. Technical Support and referrals to the users of CERCLIS database, Waste LAN and Clean LAN.

Clean LAN/Waste LAN
National .... (202) 252-0056
User support for Clean LAN, Waste LAN users.

EPA Office of Research and Development Electronic Bulletin Board (ORD BBS)
National (1200 and 2400 baud)....... (513) 569-7610
(9600 baud)............................... (513) 569-7700

STORET
National .... (800) 424-9067
.... (703) 883-8328
Technical support for STORET users.

Solid Waste Information Clearinghouse Hotline (SWICH)
National .... (800) 677-9424
....(800) 67-SWICH
Online service, modem
SWICH computer system
All aspects of solid waste management, including: source reduction, recycling, composting, planning education and training, public participation, legislation and regulation, waste combustion, collection, transfer, disposal, landfill gas and special waster.

Superfund Records of Decision System (RODS)
National .... (202) 245-3770
User support. Superfund RODs. Assists in using the RODS database, produces database reports and provides information concerning the RODS database.

Facility Index System (FINDS)
National .... (800) 424-9067
.... (703) 883-8328
Technical user support; FINDS users only.

Data Processing Support Services National Computer Center
National, except NC..................... (800) 334-9700
NC only .... (919) 541-2385
....FTS: 629-2385
User support; I.O. Control; mainframe IBM ES-9000-720 and VAX users.

Customer Technical Support National Computer Center (NCC)
National .... (800) 334-2405

NC only .... (919) 541-7862
...................................FTS: 629-7862
Provides NCC customers with technical assistance, problem diagnosis, solution and tracking.

US EPA CONTRACT LABORATORY PROGRAM (CLP)

USEPA Analytical Operations Branch (OS-230)
401 M Street, SW
Room M-2624
Washington, DC 20460
(202) 328-7906
FTS 382-7906

USEPA Contracts Mgmt. Div. (MD-33)
Alexander Drive
Research Triangle Park, NC 27111

USEPA Environmental Monitoring Systems Laboratory (EMSL/LV)
944 East Harmon Avenue
Las Vegas, NV 89108

Mailing Address:
P.O. Box 93478
Las Vegas, NV 89193-3478

Data To:
EMSL/LV Executive Center
944 East Harmon Ave.
Las Vegas, NV 89119
Attn: Data Audit Staff

USEPA National Enforcement Investigations Center (NEIC)
Denver Federal Center 53, E-2
P.O. Box 25227
Denver, CO 80225
303/236-5111 FTS 776-5111

USEPA Environmental Monitoring Systems Laboratory (EMSL/Cincinnati)
26 W. Martin Luther King Dr.
Cincinnati, OH 45268
Sample Management Office
Alexander Drive
Research Triangle Park, NC 27111

Mailing Address:
USEPA Contract Laboratory Program Sample Management Office
P.O. Box 818
Alexandria, VA 22313
(703) 557-2490
FTS 557-2490

Street Address:
Viar and Company, Inc.
300 N. Lee Street
Alexandria, VA 22314
(703) 684-5678

USEPA REGION I

USEPA Region I, WMD
J.F. Kennedy Federal Bldg.
(HPC-CAN7)
Boston, MA 02203
(617) 573-5707
FTS 833-1707

USEPA Region I, ESD
60 Westview Street
Lexington, MA 02173
(617) 860-4300
FTS 828-6300

CLP Administrative Project Officer....FTS 382-7908
CLP Technical Project Officers ...... (617) 860-4312

USEPA REGION II

USEPA Region II
26 Federal Plaza
New York, NY 10278
(212) 264-2525
FTS 340-6754

USEPA Region II, ESD
Woolbridge Avenue
Building 209
Edison, NJ 08837
(201) 321-6754
FTS 340-6754

CLP Administrative Project Officer ...FTS 382-7911
CLP Technical Project Officer..........FTS 340-6676

USEPA REGION III

USEPA Region III, Superfund Branch
841 Chestnut Street
Philadelphia, PA 19107
(215) 597-0992
FTS 597-0992

USEPA Region III, CRL
839 Bestgate Road
Annapolis, MD 21401
(301) 266-9180

CLP Administrative Project Officer ...FTS 382-7911
CLP Technical Project Officer..........FTS 266-9180

USEPA REGION IV

USEPA Region IV, Superfund Branch
345 Courtland Street, N.E.
Atlanta, GA 30365
(404) 374-4727
FTS 257-4727

USEPA Region IV, ESD (ASB)
Analytical Support Branch
College Station Road
Athens, GA 30613
(404) 546-3111
FTS 250-3111

CLP Administrative Project Officer ...FTS 382-2239
CLP Technical Project Officer..........FTS 250-3112

USEPA REGION V

USEPA Region V, ESD
536 S. Clark Street
Tenth Floor, 5SCRL
Chicago, IL 60605
FTS 353-3808

USEPA Region V, WMD
230 S. Dearborn Street
12th Floor (HR-12)
Chicago, IL 60604
(312) 886-7579
FTS 886-7579

CLP Technical Project Officer..........FTS 382-2239
CLP Technical Project Officer..........FTS 353-2313

USEPA REGION VI

USEPA Region VI Laboratory
10625 Fallstone Road
Houston, TX 77099
(713) 983-2100
FTS 730-2100

USEPA Region VI
First Interstate Tower
1445 Ross Avenue
Dallas, TX 75202
(214) 665-6491
FTS 255-6491

CLP Administrative Project Officer....FTS 382-7908
CLP Technical Project Officer ....FTS 730-2139

USEPA REGION VII

USEPA Region VII
726 Minnesota Avenue
Kansas City, KS 66101
FTS 276-7050
(913) 236-2800

USEPA Region VII
25 Funston Road
Kansas City, KS 66115
FTS 757-3881
(913) 236-3881
FAX (913) 236-2934

CLP Administrative Project Officer....FTS 382-7942
CLP Technical Project Officer..........FTS 757-3881

USEPA REGION VIII

USEPA Region VIII
999 18th Street, Suite 500
Denver, Co 80202-2405
FTS 330-1720
(303) 293-1720

Denver Federal Center
Building 5
W1 Entrance, 2nd Floor
Denver, CO 80225
FTS 776-5064
(303) 236-5073

Montana EPA Office
301 South Park
P.O. Drawer 10096
Helena, MT 59626
FTS 585-5414
(406) 449-5414

CLP Administrative Project Officer ...FTS 382-7908
CLP Technical Project Officer..........FTS 330-7511

USEPA REGION IX

USEPA Region IX, OPM, P-3-2
1235 Mission Street
San Francisco, CA 94103
FTS 556-6458
(415) 556-6458
FAX (415) 556-6874

USEPA Region IX Laboratory
944 East Harmon Avenue
Las Vegas, NV 89119

CLP Administrative Project Officer....FTS 382-7942
CLP Technical Project Officer..........FTS 556-5033

USEPA REGION X

USEPA Region X, ESD
1200 Sixth Avenue
M/S ES-095
Seattle, WA 98101
(206) 442-1200
FTS 399-1200

Laboratory USEPA Region X
P.O. Box 549
Manchester, WA 98353
(206) 442-0370
FTS 399-0370

CLP Administrative Project Officer....FTS 382-2237
CLP Technical Project Officer..........FTS 399-0370

MISCELLANEOUS INFORMATION

Cooler Returns
T. Head and Company
950 Herndon Parkway
Suite 230
Herndon, VA 22070
(703) 473-3886

ERT Edison
USEPA Environmental Response Branch
GSA Raritan Depot
Woodbridge Avenue
Edison, NJ 08837
FTS 340-6649, 6689, 6743

STATE CONTACTS

Alabama

Alabama Dept. of Environmental Management
Chief, Land Division
1751 W. L. Dickinson Drive
Montgomery, AL 36130
Phone: (205) 271-7730; Fax: (205) 271-7950

Alaska

Alaska Dept. of Environmental Conservation
Solid and Hazardous Waste Mgt.
Pouch O
Juneau, AK 99811
Phone: (907) 465-2671; Fax: (907) 789-1218

Alaska Dept. of Environmental Conservation
Chief, Spill Planning and Prevention
Pouch O
Juneau, AK 99811
Phone: (907) 465-2630; Fax: (907) 789-5673

Alaska Dept. of Environmental Conservation
Contaminated Sites Section
Pouch O
Juneau, AK 99811
Phone: (907) 465-2630; Fax: (907) 789-5072

Arizona

Arizona Dept. of Environmental Quality
Assistant Director, Office of Waste Programs
2005 N. Central Avenue, Room 701
Phoenix, AZ 85004
Phone: (602) 257-2318; Fax: (602) 257-6874

Arizona Dept. of Environmental Quality
Mgr., Waste Assessment Section
2005 N. Central, Room 403C
Phoenix, AZ 85004
Phone: (602) 257-6995; Fax: (602) 257-6948

Arkansas

Arkansas Dept. of Pollution Control and Ecology
Chief, Hazardous Waste Division
P.O. Box 8913
Little Rock, AR 72219-8913
Phone: (501) 562-7444 or (501) 562-6533; Fax:
(501) 562-4632 or (501) 562-2541

California

California Dept. of Toxic Substances Control
Chief, Program and Administrative Support Div.
P.O. Box 806
Sacramento, CA 95812-0806
Phone: (916) 324-7193
Fax: (916) 324-1788

California State Water Resources Control Board
Chief, Div. of Clean Water Programs
P.O. Box 944212
Sacramento, CA 94244-2120
Phone: (916) 739-4333; Fax: (916) 739-2300

Colorado

Colorado Dept. of Health
Hazardous Materials and Waste Mgt. Div.
4210 E. 11th Avenue
Denver, CO 80220
Phone: (303) 331-4830; Fax: (303) 331-4401

Public Utilities Commission
Hazardous Materials Transportation Permits
1580 Logan Street, Off. Level 1
Denver, CO 80203
Phone: (303) 894-2010; Fax: (303) 894-2065

Connecticut

Connecticut Waste Management Bureau
Bureau Chief
State Office Building/165 Capitol Ave.
Hartford, CT 06106
Phone: (203) 566-8476; Fax: (203) 566-4924

Connecticut Resource Recovery Authority
President
179 Allyn St., Suite 603
Hartford, CT 06103
Phone: (203) 549-6390; Fax: (203) 522-2390

Delaware

Delaware Dept. of Natural Resources and
    Environmental Control
Mgr., Hazardous Waste Management Branch
P.O. Box 1401/89 Kings Highway
Dover, DE 19903
Phone: (302) 739-3689; Fax: (302) 739-5060

District of Columbia

Dept. of Consumer and Regulatory Affairs (DCRA)
Chief, Pesticides and Hazardous Waste Mgt.
    Branch
2100 Martin Luther King, Jr. Ave., SE Suite #203
Washington, DC 20020
Phone: (202) 404-1167; Fax: (202) 404-1188

Florida

Florida Dept. of Environmental Regulation
Administrator, Solid and Haz. Waste
Underground Storage Tanks
Twin Towers Office Bldg
2600 Blair Stone Road
Tallahassee, FL 32399-2400
Phone: (904) 488-0300; Fax: (904) 922-4939

Georgia

Land Protection Branch
Chief
Floyd Towers East
205 Butler Street, SE
Atlanta, GA 30334
Phone: (404) 656-2833; Fax: (404) 651-9425

Hawaii

Hawaii Dept. of Health
Mgr, Solid and Hazardous Waste Branch
Five Waterfront Plaza, Suite 250
500 Ala Moana Boulevard
Honolulu, HI 96813
Phone: (808) 543-8225; Fax: (808) 548-7237

Hawaii Dept. of Health
Mgr., Hazard Evaluation and Emerg. Response
Five Waterfront Plaza, Suite 250
500 Ala Moana Boulevard
Honolulu, HI 96813
Phone: (808) 543-8248; Fax: (808) 548-7237

Idaho

Idaho Dept. of Health and Welfare
Mgr., RCRA Programs
1410 North Hilton Street
Boise, ID 83706
Phone: (208) 334-5879; Fax: (208) 334-0417

Illinois

Illinois Environmental Protection Agency
Director
2200 Churchill Road
Springfield, IL 62706
Phone: (217) 782-3397; Fax: (217) 782-9039

Illinois Environmental Protection Agency
Public Information Officer
Division of Land Pollution Control
2200 Churchill Road
Springfield, IL 62706
Phone: (217) 782-6760; Fax: (217) 524-4193

Hazardous Waste Research and Information Center
Illinois Energy and Natural Resources
David Thomas - Director
1 E. Hazelwood Drive
Champaign, IL 61820
Phone: (217) 333-8941; Fax: (217) 333-8944

Indiana

Indiana Dept. of Environmental Management
Branch Chief, Office of Hazardous Waste Mgt.
105 S. Meridian St./P.O. Box 6015
Indianapolis, IN 46225
Phone: (317) 232-3292; Fax: (317) 232-3403

Kansas

Bureau of Air and Waste Management
Kansas Dept. of Health and Environment
Director
Forbes Field, Building 740
Topeka, KS 66620
Phone: (913) 296-1593; Fax: (913) 296-6247

Kentucky

Kentucky Dept. of Environmental Protection
Director, Div. of Waste Mgt.
Omega Bldg., Ft. Boone Plaza
Frankfort, KY 40601

Phone: (502) 564-6716 ext 214;
Fax: (502) 564-4245

Louisiana

Louisiana Dept. of Environmental Quality
Office of Solid and Hazardous Waste
P.O. Box 82178
Baton Rouge, LA 70884-2178
Phone: (504) 765-0355; Fax: (504) 765-0617

Louisiana Dept. of Environmental Quality
Administrator, Ground Water Protection Division
P.O. Box 82215
Baton Rouge, LA 70884-2215
Phone: (504) 765-0585;

Maine

Bureau of Hazardous Mat'ls and Solid Waste
  Control
Director - Maine Dept of Environmental Protection
State House Station #17
Augusta, ME 04333; Phone: (207) 289-2651

Maryland

Hazardous and Solid Waste Management Admin.
Maryland Dept. of the Environment
Director
2500 Broening Highway
Baltimore, MD 21224
Phone: (301) 631-3304; Fax: (301) 631-3321

Massachusetts

Massachusetts Dept. of Environmental Affairs
Director, Executive Office
100 Cambridge Street, 20th Floor
Boston, MA 02202
Phone: (617) 727-9800; Fax: (617) 727-2754

Michigan

Waste Management Division
Michigan Dept. of Natural Resources
Chief, Hazardous Waste Permits Section
P.O. Box 30241
Lansing, MI 48909
Phone: (517) 373-0503; Fax: (517) 373-4797

Minnesota

Minnesota Pollution Control Agency
Director, Hazardous Waste Div.
520 Lafayette Rd. North
St. Paul, MN 55155
Phone: (612) 297-8502; Fax: (612) 297-8676

Hazardous Waste Division
Minnesota Pollution Control Agency
Chief, Program Development
520 Lafayette Rd. North
St. Paul, MN 55155
Phone: (612) 297-8355; Fax: (612) 297-8676

Minnesota Technical Assistance Program (MnTAP)
Director
1313 5th Street, SE - Suite 207
Minneapolis, MN 55414
Phone: (612) 627-4646 or (800) 247-0015;
Fax: (612) 627-4769

Mississippi

Mississippi Dept. of Environmental Quality
Chief, Hazardous Waste Division
P.O. Box 10385
Jackson, MS 39289-0385
Phone: (601) 961-5062; Fax: (601) 961-5062

Missouri

Missouri Dept. of Natural Resources
Dir., Hazardous Waste Program
205 Jefferson St./P.O. Box 176
Jefferson City, MO 65102
Phone: (314) 751-3176; Fax: (314) 751-7869

Montana

Montana Dept. of Health and Environmental
   Sciences
Chief, Solid and Hazardous Waste Bureau
836 Front Street
Helena, MT 59620
Phone: (406) 444-2821; Fax: (406) 444-1499

Nebraska

Nebraska Dept. of Environmental Control
Hazardous Waste Sec., CERCLA Unit Supervisor
P.O. Box 98922
Lincoln, NE 68509
Phone: (402) 471-2186

Nebraska Dept. of Environmental Control
Hazardous Waste Sec., RCRA Unit Supervisor
P.O. Box 98922
Lincoln, NE 68509
Phone: (402) 471-2186; Fax: (402) 471-2909

Nevada

Division of Environmental Protection
Nevada Dept. of Conserv. and Natural Resources
Chief, Waste Mgt. Bureau
Capitol Complex
123 W. Nye Lane
Carson City, NV 89710
Phone: (702) 687-5872; Fax: (702) 885-0868

New Hampshire

New Hampshire Dept. of Environmental Services
Director, Waste Mgt. Division
6 Haven Drive
Concord, NH 03301-6509
Phone: (603) 271-2906; Fax: (603) 271-2867

New Jersey

New Jersey Dept. of Environmental Protection
Assistant Commissioner, Site Remediation Program

401 E. State Street, CN 028-6th Floor East
Trenton, NJ 08625
Phone: (609) 292-1250; Fax: (609) 633-2360

New Jersey Dept. of Environmental Protection
   and Energy
Dir., Div. of Responsible Party Site Remediation
401 E. State Street, CN 028-5th Floor East
Trenton, NJ 08625
Phone: (609) 633-1408; Fax: (609) 633-1454

New Mexico

New Mexico Environment Dept.
Chief, Hazardous and Radioactive Mat'ls Bureau
P.O. Box 26110
1190 St. Francis Drive
Santa Fe, NM 88002
Phone: (505) 827-2922; Fax: (505) 827-0581

New York

New York State Dept. of Environmental Conserv.
Dir., Div. of Hazardous Substances Regulation
50 Wolf Rd., Room 229
Albany, NY 12233-7250
Phone: (518) 457-6934; Fax: (518) 457-0629

North Carolina

N. Carolina Dept. of Environmental, Health,
   and Natural Resources
Dir., Div. of Solid Waste Mgt.
P.O. Box 27687
Raleigh, NC 27611-7687
Phone: (919) 733-4996; Fax: (919) 733-4810

North Dakota

N. Dakota Dept. of Health and Consolidated Labs
Div. of Waste Mgt.
1200 Missouri Ave, Rm. 302
Bismarck, ND 58502-5520
Phone: (701) 221-5166; Fax: (701) 221-5200

Ohio

Ohio Environmental Protection Agency
Chief, Div. of Solid and Hazardous Waste Mgt.
1800 Watermark Drive
P.O. Box 1049
Columbus, OH 43266-0149
Phone: (614) 644-2917; Fax: (614) 644-2329

Oklahoma

Oklahoma State Dept. of Health
Chief, Hazardous Waste Mgt. Service - 0205
1000 Northeast Tenth Street
Oklahoma City, OK 73117-1299
Phone: (405) 271-5338; Fax: (405) 271-5205

Oregon

Oregon Dept. of Environmental Quality
Administrator, Hazardous and Solid Waste Div.
811 SW 6th Ave.
Portland, OR 97204
Phone: (503) 229-5356; Fax: (503) 229-6124

Oregon Dept. of Environmental Quality
Administrator, Environmental Cleanup Div.
811 SW 6th Ave.
Portland, OR 97204
Phone: (503) 229-5254; Fax: (503) 229-6124

Pennsylvania

Pennsylvania Dept. of Environmental Resources
Dir., Bureau of Waste Mgt.
P.O. Box 2063
Harrisburg, PA 17105-2063
Phone: (717) 783-2388; Fax: (717) 787-1904

Pennsylvania Dept. of Environmental Resources
Chief, Division of Permits
P.O. Box 2063
Harrisburg, PA 17105-2063
Phone: (717) 787-7381; Fax: (717) 787-1904

Rhode Island

Rhode Island Dept. of Environmental Mgt.
Dir., Div. of Air and Hazardous Materials
291 Promenade Street
Providence, RI 02908
Phone: (401) 277-2797

South Carolina

S. Carolina Dept. of Health and Environmental
 Control
Chief, Bureau of Solid and Hazardous Waste Mgt.
2600 Bull St.
Columbia, SC 29400
Phone: (803) 734-5164; Fax: (803) 734-5199

South Dakota

S. Dakota Dept. of Environment and Natural
 Resources
Office of Waste Mgt.
319 S. Coteau c/o 500 E. Capital Avenue
Pierre, SD 57501
Phone: (605) 773-3105; Fax: (605) 773-6046

S. Dakota Highway Patrol, Commerce and
 Regulation
320 N. Nicollet
Pierre, SD 57501
Phone: (605) 773-3105; Fax: (605) 773-6046

Tennessee

Tennessee Dept. of Environment and Conservation
Dir., Div. of Solid Waste Mgt.
701 Broadway
Customs House, 4th Floor
Nashville, TN 37243-1535
Phone: (615) 741-3424; Fax: (615) 741-4666

Tennessee Dept. of Environment and Conservation
Dir., Div. of Superfund
Doctors Bldg, 706 Church Street
Nashville, TN 37243-1538
Phone: (615) 741-6287

Texas

Texas Water Commission
Dir., Hazardous and Solid Waste Division
P.O. Box 13087, Capitol Station
Austin, TX 78711-3087
Phone: (512) 463-7760; Fax: (512) 463-8408
Texas Dept. of Health
Occupational Safety and Health Division
1100 West 49th Street
Austin, TX 78756-3199
Phone: (512) 459-1611

Utah

Utah Dept. of Environmental Quality
Dir., Div. of Solid and Hazardous Waste
288 North 1460 West St.
Salt Lake City, UT 84114-4880
Phone: (801) 538-6170; Fax: (801) 538-6016

Vermont

Vermont Agency of Natural Resources
Dir., Hazardous Mat'ls Mgt. Div.
103 South Main St.
Waterbury, VT 05676
Phone: (802) 244-8702; Fax: (802) 244-5141

Vermont Dept. of Health
Dir., Occupations and Radiological Health Div.
10 Baldwin Street
Montpelier, VT 05642
Phone: (802) 828-2886; Fax: (802) 828-2878

Virginia

Virginia Dept. of Waste Management
Div. of Regulation
Monroe Building, 11th Floor
1201 N 14th St.
Richmond, VA 23219
Phone: (804) 225-2667; Fax: (804) 225-3753

Washington

Washington Dept. of Ecology
Mgr., Solid and Hazardous Waste Program
Mail Stop PV-II
Olympia, WA 98504
Phone: (206) 459-6316; Fax: (206) 438-7759

West Virginia

Division of Natural Resources
West Virginia Dept. of Commerce, Labor and
 Environmental Resources
Chief, Waste Mgt. Section
1356 Hansford St.
Charleston, WV 25301
Phone: (304) 348-5929; Fax: (304) 348-0256

Air Pollution Control Commission
Director
1558 Washington Street, East
Charleston, WV 25311
Phone: (304) 348-2275

West Virginia Division of Highways
Acting Commissioner, State Highway Engineer
Building 5, Room A-125
Charleston, WV 25305
Phone: (304) 348-3505

Wisconsin

Wisconsin Dept. of Natural Resources
Director, Bureau of Solid and Hazardous Waste
    Mgt.
P.O. Box 7921
Madison, WI 53707
Phone: (608) 266-1327; Fax: (608) 267-2768

Wyoming

Wyoming Dept. of Environmental Quality
Supervisor, Solid Waste Mgt. Program
122 West 25th St.
Herschler Building
Cheyenne, WY 82002
Phone: (307) 777-7752; Fax: (307) 777-5973

APPENDIX F

Source information used to compile this manual is the latest promulgated versions of methods commonly available to analytical testing laboratories and environmental engineering companies. In some cases newer unpromulgated versions of methods exist; these are not included in this reference because there use is not universally accepted. The latest versions of the inorganic and organic Statements of Work for the Contract Laboratory Program were used, though some laboratories may still have contracts using an older Statement of Work version. SW-846 third edition, Final revision 1 methods were promulgated in July 1992 and are included in this reference.

REFERENCES

U.S. Environmental Protection Agency
National Primary Drinking Water Regulations. 40 CFR Part 141; July 1, 1988.

Contract Laboratory Program – User's Guide to the Contract Laboratory Program. EPA/540/P-91/002, January 1991, Office of Emergency and Remedial Response. Washington, DC

Contract Laboratory Program – Statement of Work for Organic Analysis, Multi-Media, Multi-Concentration. Document OLMO1.0 – OLMO1.9, July 1993

Contract Laboratory Program - Statement of Work for Inorganic Analysis – Multi-Media Multi-Concentration. Document ILMO2.0 – ILMO3.0, 1992

EPA100-400 Series – Methods for Chemical Analysis of Water and Wastes, EPA-600/4-79-020, March 1983.

EPA 500 Series – Methods for the Determination of Organic Compounds in Drinking Water, EPA/600/4-88/039, December 1988.

EPA 500 Series – Methods for the Determination of Organic Compounds in Drinking Water, Supplement I, EPA/600/4-90/020, 1990.

EPA 600 Series – 40 CFR, Part 136, July 1, 1988.

Variability in Protocols – Guy F. Simes, Risk Reduction Engineering Laboratory, Cincinnati, OH, September 1991

Test Methods for Evaluating Solid Waste Physical/Chemical
Second Edition Rev 0, September 1986, and Third Edition, Final Update 1, July 1992. Includes all updates and corrections through July 1992.

Standard Methods for the Examination of Water and Waste Water.
18th Edition 1992. American Public Health Association, American Water Works Association, Water Pollution Control Federation.

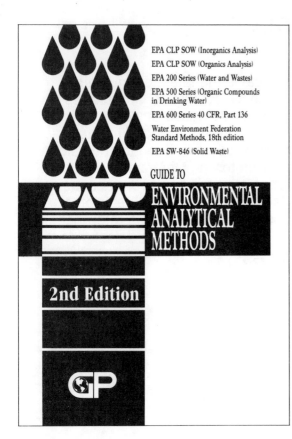

EPA CLP SOW (Inorganics Analysis)
EPA CLP SOW (Organics Analysis)
EPA 200 Series (Water and Wastes)
EPA 500 Series (Organic Compounds in Drinking Water)
EPA 600 Series 40 CFR, Part 136
Water Environment Federation Standard Methods, 18th edition
EPA SW-846 (Solid Waste)

GUIDE TO

ENVIRONMENTAL ANALYTICAL METHODS

2nd Edition

GP

Prices for second edition:

1 copy - **$34.05**; 5 copies - **$24.65** each; 10 copies - **$19.10** each; 25 copies - **$17.05** each; 50 copies - **$15.30** each; 100 copies - **$11.50** each

Need more copies?

If your associates are clamoring for more copies of this book, or your own copy seems to be constantly disappearing, it's easy to obtain more. Simply direct your order to Genium at the address below. A purchase order or payment must accompany all orders. There are no shipping charges when payment accompanies order. Shipping on unpaid orders is as follows: $6 on orders of less than $100; $8 on orders from $100 - $299; $15 on orders from 300 to $1,000; and actual freight cost on orders exceeding $1,000. New York State residents, please add appropriate state and local sales tax. All payments must be in U.S. dollars. Please make check payable to Genium Publishing Corporation.

Imprinting is also available!

Imprinting makes the *Guide to Environmental Analytical Methods* a great handout to customers and employees alike. Suppliers to the environmental laboratory industry have found the *Guide* to be an appreciated premium to existing and potential customers. Genium will custom imprint the front cover of your books free of charge on orders of 50 copies or more. The maximum imprint area is 3 1/2 inches wide by 4 inches deep. Genium will typeset and layout your message at no additional charge. If you want your organization's logo or your own typeset copy included with your imprint, please send camera-ready artwork (a clear, crisp black-and-white reproduction - no photocopies or business cards). We can also add your camera-ready logo to the copy we typeset. A one-time $35.00 printing plate charge applies to all orders with logos or artwork. This extra plate charge is waived on reorders.

Please direct all orders and inquiries to:

GENIUM PUBLISHING CORPORATION
Room 27; One Genium Plaza
Schenectady, NY 12304-4690 USA
Toll Free: (800) 2-GENIUM
Telephone: (518) 377-8854
Fax: (518) 377-1891